Systems Simulation and Modeling for Cloud Computing and Big Data Applications

Advances in Ubiquitous Sensing
Applications for Healthcare

Systems Simulation and Modeling for Cloud Computing and Big Data Applications

Volume Ten

Series Editors

Nilanjan Dey

Amira S. Ashour

Simon James Fong

Volume Editors

J. Dinesh Peter
*Karunya Institute of Technology and Sciences,
Coimbatore, Tamilnadu, India*

Steven L. Fernandes
*University of Alabama at Birmingham, Birmingham, AL,
United States*

ACADEMIC PRESS
An imprint of Elsevier

ELSEVIER

Academic Press is an imprint of Elsevier
125 London Wall, London EC2Y 5AS, United Kingdom
525 B Street, Suite 1650, San Diego, CA 92101, United States
50 Hampshire Street, 5th Floor, Cambridge, MA 02139, United States
The Boulevard, Langford Lane, Kidlington, Oxford OX5 1GB, United Kingdom

Notices
Knowledge and best practice in this field are constantly changing. As new research and experience broaden our understanding, changes in research methods, professional practices, or medical treatment may become necessary.

Practitioners and researchers must always rely on their own experience and knowledge in evaluating and using any information, methods, compounds, or experiments described herein. In using such information or methods they should be mindful of their own safety and the safety of others, including parties for whom they have a professional responsibility.

To the fullest extent of the law, neither the Publisher nor the authors, contributors, or editors, assume any liability for any injury and/or damage to persons or property as a matter of products liability, negligence or otherwise, or from any use or operation of any methods, products, instructions, or ideas contained in the material herein.

Library of Congress Cataloging-in-Publication Data
A catalog record for this book is available from the Library of Congress

British Library Cataloguing-in-Publication Data
A catalogue record for this book is available from the British Library

ISBN 978-0-12-819779-0

For information on all Academic Press publications
visit our website at https://www.elsevier.com/books-and-journals

Publisher: Mara Conner
Acquisitions Editor: Chris Katsaropoulos
Editorial Project Manager: Ali Afzal-Khan
Production Project Manager: Punithavathy Govindaradjane
Cover Designer: Matthew Limbert

Typeset by SPi Global, India

Working together
to grow libraries in
developing countries

www.elsevier.com • www.bookaid.org

Contents

CHAPTER 6 Predictive analysis of diabetic women patients using R **99**

R. Rifat Ameena and B. Ashadevi

CHAPTER 7 IoT-based smart mirror for health monitoring **115**

Ilango Krishnamurthy, D. Prabha, and M.S. Karthika

CHAPTER 8   Discovering human influenza virus using ensemble learning **123**

M. Nandhini and M.S. Vijaya

CHAPTER 9   Mining and monitoring human activity patterns in smart environment-based healthcare systems **137**

M. Janani, M. Nataraj, and C.R. Shyam Ganesh

Contributors

S. Akila Agnes
Department of Computer Science and Engineering, Karunya Institute of Technology and Sciences, Coimbatore, India

R. Rifat Ameena
Department of Computer Science, M.V. Muthiah Government Arts College for Women, Dindigul, India

S. Meenakshi Ammal
Department of CSE, PSG College of Technology, Coimbatore, India

J. Anitha
Department of Computer Science and Engineering, Karunya Institute of Technology and Sciences, Coimbatore, India

Diana Arulkumar
Department of Computer Science and Engineering, Karunya Institute of Technology and Sciences, Coimbatore, India

B. Ashadevi
Department of Computer Science, M.V. Muthiah Government Arts College for Women, Dindigul, India

R.V. Belfin
Department of Computer Science and Engineering, Karunya Institute of Technology and Sciences, Coimbatore, India

Tonmoay Deb
Department of Electrical and Computer Engineering, North South University, Dhaka, Bangladesh

Suganthi Evangeline
Electronics and Communication Engineering Department, Karunya Institute of Technology and Science, Coimbatore, India

Adnan Firoze
Department of Electrical and Computer Engineering, North South University, Dhaka, Bangladesh

C.R. Shyam Ganesh
Department of CSE, PSG College of Technology, Coimbatore, India

Deva Priya Isravel
Department of Computer Science and Engineering, Karunya Institute of Technology and Sciences, Coimbatore, India

Biju Issac
Department of Computer and Information Sciences, Northumbria University, Newcastle, United Kingdom

Titus Issac
Karunya Institute of Technology and Sciences, Coimbatore, India

M. Janani
Department of CSE, Jai Shriram Engineering College, Tirupur, India

L.S. Jayashree
Department of CSE, PSG College of Technology, Coimbatore, India

M.S. Karthika
Information Technology, Bannari Amman Institute of Technology, Sathyamangalam, Coimbatore, India

Ilango Krishnamurthy
Computer Science and Engineering, Sri Krishna College of Engineering and Technology, Coimbatore, India

X. Anitha Mary
Instrumentation Engineering Department, Karunya Institute of Technology and Science, Coimbatore, India

Shoba Mohan
Holoteq Group, Doha, Qatar

M. Nandhini
PSGR Krishnammal College for Women, Coimbatore, India

M. Nataraj
Department of CSE, Kongu Engineering College, Erode, India

Tousif Osman
Department of Electrical and Computer Engineering, North South University, Dhaka, Bangladesh

S. Immanuel Alex Pandian
Department of Electronic and Communication Engineering, Karunya Institute of Technology and Sciences, Coimbatore, India

D. Prabha
Computer Science and Engineering, Sri Krishna College of Engineering and Technology, Coimbatore, India

Shahreen Shahjahan Psyche
Department of Electrical and Computer Engineering, North South University, Dhaka, Bangladesh

Rashedur M. Rahman
Department of Electrical and Computer Engineering, North South University, Dhaka, Bangladesh

Kumudha Raimond
Department of Computer Science and Engineering, Karunya Institute of Technology and Sciences, Coimbatore, India

K. Rajasekaran
Instrumentation Engineering Department, Karunya Institute of Technology and Science, Coimbatore, India

Elijah Blessing Rajsingh
Karunya Institute of Technology and Sciences, Coimbatore, India

Salaja Silas
Karunya Institute of Technology and Sciences, Coimbatore, India

I-Hsien Ting
Department of Information Management, National University of Kaohsiung, Kaohsiung, Taiwan

M.S. Vijaya
PSGR Krishnammal College for Women, Coimbatore, India

Preface

Systems Simulation and Modeling (SSM) is a discipline that focuses on solving problems through the use of models and simulations. SSM is used in almost every science and engineering involving multidisciplinary research. SSM develops frameworks that are applicable across disciplines to develop benchmark tools that are useful in solution development. For SSM to grow and continue to develop, modeling theories need to be transformed into consistent frameworks, which in turn are implemented into consistent benchmarks. The world is clearly in the era of big data and cloud computing. The challenge for big data is balancing operation and cost tradeoffs by optimizing configuration at both the hardware and software layers to accommodate users' constraints. Conducting such a study in a real-time computing environment can be difficult for the following reasons:

- establishing or renting a large-scale datacenter resource pool
- frequently changing experiment configurations in a large-scale real workplace involves significant manual configuration
- comprising and controlling different types of failure behaviors and benchmarks across heterogeneous software and hardware resource types in a real workplace

SSM-based approaches to performance testing and benchmarking offer significant advantages. Many big data, cloud application developers and researchers can perform tests in a controllable and repeatable manner. Fired by the need to analyze the performance of different big data processing and cloud frameworks, researchers have introduced several benchmarks, including BigDataBench, BigBench, HiBench, PigMix, CloudSuite, and GridMix. Although substantial progress has been made, the research community still needs a holistic comprehensive big data and cloud simulation platform for different applications.

The editors appreciate the extensive time and effort put in by all the chapter contributors, which has ensured a high quality of content. The editors would also like to express their thanks to the panel of experts who assisted in reviewing the chapters.

Differential color harmony: A robust approach for extracting harmonic color features and perceiving aesthetics in a large image dataset

1

**Tousif Osman, Shahreen Shahjahan Psyche, Tonmoay Deb, Adnan Firoze,
Rashedur M. Rahman**

Department of Electrical and Computer Engineering, North South University, Dhaka, Bangladesh

1.1 Introduction

Understanding beauty and aesthetics has always been of interest for humans, from the very beginnings of human history. Many artists through the ages have invented their own styles of capturing beauty and have produced artwork that has dazzled, and continues to dazzle, mankind to this day. One of the difficulties of working in this domain is that there is no exact measurement of beauty as humans perceive it. Many artists have formulated their own idea of aesthetics, but none of them are absolute nor do they involve an exact scale of measurement. Therefore, most regular mathematics and computational techniques fail in this domain.

The idea of color harmony and its correlation with aesthetics is not new. One of the oldest and most notable works in this area [1], which had a great impact, was by Moon and Spencer in 1944. Many ideas and theories came after that work and today this field is quite enriched. The concept of color harmony revolves around the idea that colors maintaining a certain relation with their neighboring colors are perceived to be more appealing to the human eye [2]. This concept is very similar to the concept of musical notation, and is one of the few laws of art that does not vary from person to person. Currently, one of the most commonly used basic ideas for extracting harmonic color features (HCFs) is clustering and segmenting the colors. However, this is a resource-hungry, soft-computational approach.

In our research, we have used mathematical and computational approaches to provide a solution to this problem. We have used neural networks (NNs) and regression models to verify our extracted features and have demonstrated their significance in real-world situations. Even with simpler regression models, i.e., linear regression, the results are still promising. The procedures and workflow of our algorithm have been described chronologically in the following sections.

1.2 Related works

We are not the only ones who have considered color harmony in measuring the aesthetic beauty of an image. Other researchers have walked this path before us. Lu et al. [3] have designed a statistical learning structure to train a color harmony model. They used a dataset consisting of a large number of natural images. They made Dirichlet allocation training (LDA) smoother by using the content of the images along with the visual features. Then, the harmonic color level was estimated based on the supervised/unsupervised model(s) that indicate the photo aesthetic scores. In another paper, Lu et al. [4] trained an LDA-based color harmony model that considers harmonic colors as using spatial distances.

Datta et al. [5] have developed a system that automatically infers the aesthetic beauty of an image using the visual content. They crowd sourced the ratings of images. They employed classifiers using machine-learning techniques, such as support vector machine, classification trees, and linear regression, to predict a numerical aesthetic score.

Phan et al. [6] have built statistical models that are even used in building some practical applications. These statistical models have been trained based on the coloring style of a fine arts collection. The authors have also used density estimation to determine the features of palette data, since artists usually have their own personal coloring patterns in their creations, which result in the frequent appearance of certain color palettes in multiple paintings of the same artist.

Amati et al. [7] have shown that there is no explicit linear dependence between colorfulness and aesthetics; rather, correlations arise categorically for different images: for example, "landscape," "abstract." As a dataset, they have compiled perceptual data from a large-scale user study.

Lu et al. [8] have proposed a color harmony model based on a Bayesian framework. Classical artists had already paved way using this technique unconsciously, without adhering to any established framework. On the other hand, learning-based models that detect the underlying patterns of a work are shown to be a possibility. In the LDA learning process, other two-color harmony models have been used during the training of the model.

Nishiyama et al. [9] claim that the existing harmonic color models consider simple coloring patterns that fail to assess photographs with complicated color patterns. To resolve this challenging problem, they have built a method in which they consider a photograph consisting of a cluster of confined regions with variation in colors that are simpler, which has ultimately led them towards developing a method to assess the aesthetics of a photo based on its harmonious colors. They have also improved the classification performance by integrating blur, edges, and saliency features.

1.3 **Methodology**

The basic principle of our research stands on the following idea: colors seem to be harmonized to the human eye if the color shades (Fig. 1.1A) or the base color changes (Fig. 1.1B) gradually maintain a pattern [2]. Fig. 1.1 shows three combinations of colors: (A) gradual change of color shade, (B) gradual change of color base and shade, and (C) no pattern in the change of colors. Looking at these three images, even a nonartist will find Fig. 1.1A and B more appealing than Fig. 1.1C. By nature, the human eye tends to find harmonic color "melodies" similar to music.

Commonly, artists and designers shift colors from RGB (red, green, blue) color scale to HSV/HSL (hue, saturation, value/hue, saturation, lightness). For our research, we selected the HSV scale. The RGB color scale states that every color can be represented with the three base color components of red, green and blue. In the case of HSV, in simple terms, the hue is the key component that defines the base of a color, saturation defines the whiteness of a color, and value specifies the blackness of a color. This color scale represents all possible colors in terms of these three components.

According to designer and artist guidelines, a small change in the HSV/HSL color component produces a color that maintains harmony with the original color. We have used these facts as the base hypothesis of our research. Fig. 1.2 shows a three-dimensional representation of RGB (Fig. 1.2A) and HSV (Fig. 1.2B) color space, respectively.

(A)   (B)   (C)

FIG. 1.1

Scale of differential colors. (A) Harmonious shade, (B) harmonious hue, (C) random color combination.

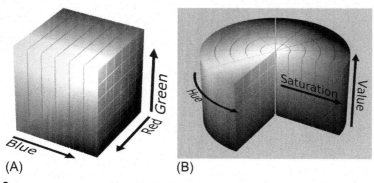

(A)   (B)

FIG. 1.2

Three-dimensional representation of color spaces. (A) RGB [10], (B) HSV [10].

After shifting space, we then calculated the gradient of the shifted color space. This gradient tells us about the rate of change in color components. Next, we applied differential operators to produce a smaller feature set that tells about the overall color correlation in an image. Finally, we applied a machine-learning model. Details of the feature extraction process and model design are explained in the following subsections.

1.3.1 Extracting differential harmonic color features

This is the initial and most crucial stage of our workflow and each step has been ordered based on performance and results (of the previous step). Shifting color spaces from RGB to HSV is the very first step of our process. We are shifting the color space so that a linear change in color component represents a change in color shade or base. This is also a common process in much color analysis research [3, 5, 8], to make the colors more aligned with how humans perceive colors. Fig. 1.3A is a sample image and Fig. 1.3B–D are its three RGB color components, respectively. Fig. 1.3E–G are the components of shifted HSV components.

Next, to produce a fixed-length feature vector, we have transformed the variable size images to a 128-pixel by 128-pixel square image using interpolation. Our dataset (described in more detail in Section 1.3.3) keeps the images in the range of 128×128, but they are not square images: an image can be 128×100 or of some other dimensions. For regular images, the size can vary in a very large domain. Therefore, if we do not change the size to a fixed-size image, the final features will not have a fixed length. We have used bicubic interpolation [11]. Fig. 1.4 shows the transformation of the sample image.

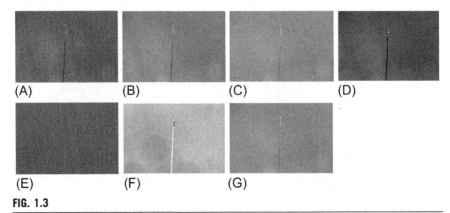

FIG. 1.3

One input image and components. (A) Input image, (B) red channel, (C) green channel, (D) blue channel, (E) hue channel in HSV, (F) saturation channel in HSV, and (G) value channel of HSV.

FIG. 1.4

Resized/resampled image version of 128 × 128 pixels from Fig. 1.3A.

The change of color component rate is formulated using a gradient image by computing the average distance of each pixel with its surrounding eight pixels. Eq. (1.1) has been applied to each pixel of the components to produce the gradient images:

$$g(x, y) = \sum_{i,j=-1,-1}^{2,2\, i,j\neq 0,0} \frac{cmp(x, y) - cmp(x+i, y+j)}{8} \tag{1.1}$$

Here x and y are the iterators of each component and cmp is the variable representing the component matrix. Fig. 1.5 shows the gradients of each component. Fig. 1.5A illustrates the gradient image and Fig. 1.5B–D illustrates the HSV gradient components, respectively.

Following is the pseudocode for producing gradient images:

```
For p_ij in Image Component:
  For i from −1 to 2
    For j from −1 to 2
      If i, j is not (0,0)
        avg_dist := avg_dist + (comp[p_ij] − comp[p_i + i, p_j + j])/8
    End if
  End for
  End for
  gra_comp[p_ij] := avg_dist
End for
```

Here, p_{ij} is an iterator on the color component and the ij suffix represents the column and row iterator variable. The gra_comp is the final gradient of the component for the given component. Resulting gradients hold the information about color change, but

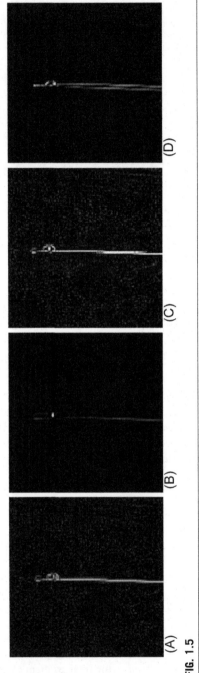

FIG. 1.5

Gradients of each color component. (A) Combined gradient, (B) hue gradient, (C) saturation gradient, (D) value gradient.

due to their higher dimensions they are problematic as features. The feature vector has a length of $3 \times 16{,}384$ and moreover we have 5000 sample images. Regular learning algorithms will fail to map the features and will require a long time to process without dimensionality reduction. We could apply principal component analysis (PCA) directly on the entire dataset, but by applying derivative and min-max normalization, we are extracting a robust feature set that has the information about the change of color in an image without much loss and can be used in any algorithm directly. However, by doing differentiation, we are losing the detail information, but we are only interested in the rate of change and continued differentiation is preserving it without much loss.

From our research results, we have concluded that applying discrete differentiation on the gradients holds the color correlation for the entire image from a lower dimension. Eq. (1.2) is the discrete differentiation operator we have used to reduce the dimensions. $X_1, X_2, X_3, X_{m-1}, X_m$ are the elements of the feature vector, where X_1 is the first element and X_m is the last element.

$$\Delta y = [X_2 - X_1, X_3 - X_2, \ldots X_m - X_{m-1}] \tag{1.2}$$

This operator reduces one dimension of a series. We have calculated lower dimensional 5×5 gradient matrices by calculating 123rd derivatives and subsequently calculating 123rd derivatives on the transpose of the previously calculated matrix.

$$z_i = \frac{x_i - \min(x)}{\max(x) - \min(x)} \tag{1.3}$$

We have applied min-max normalization to normalize the results in Eq. (1.2) (using Eq. 1.3). Fig. 1.6A–C represent differential matrices derived from each component.

Finally, we have changed the shape 5×5 2D matrix to a 1×25 1D matrix. These final matrices are our calculated HCF vectors, which we used to train models in the following sections. Fig. 1.7A–C represents the nonnormalized feature scores. If we observe the original sample image (Fig. 1.4), we can see the image has a base color of green, where the greatest part of the image is composed of shades of green (light gray in print version), which implies the rate of change of hue is less, whereas the saturation and value have a higher rate of change. Our calculated HCF also complies with the expected result: hue has a more stable rate of change than the other two. Fig. 1.8A

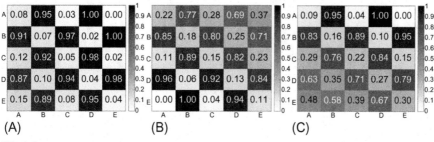

(A)   (B)   (C)

FIG. 1.6

Differential features. (A) Hue, (B) saturation, and (C) value component.

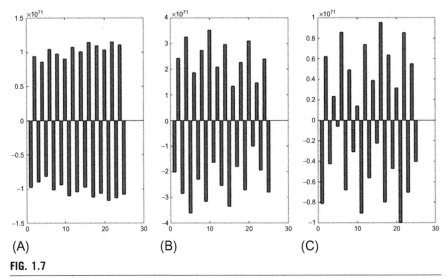

FIG. 1.7

Harmonic color features (HCF) plots: (A) hue, (B) saturation, (C) value.

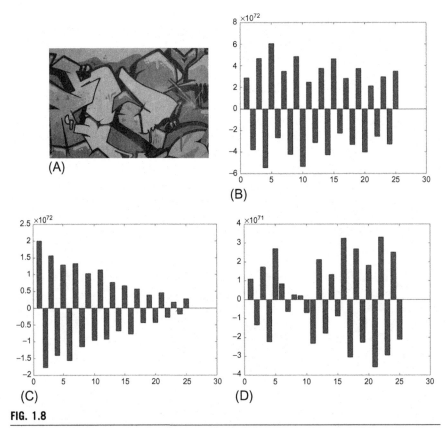

FIG. 1.8

Input image and harmonic color features (HCF) plots. (A) Input image, (B) hue, (C) saturation, (D) value.

represents another sample image where color in the image changes more rapidly. Fig. 1.8B–D represent their differential feature components.

1.3.2 Regression model design

We have applied our feature vectors on a real-world dataset and predicted an aesthetics score using these features. A multilayer NN model was used to train our system. We constructed one hidden layer with a hidden layer size of 5 in our NN. We trained the model with .45 learning rate and 500 training cycles. Combining three components, we used 75 features as our color features. We have preprocessed the feature set using PCA and validated our results using bootstrapping. We have taken 7 PCAs that capture 95% of variance. Fig. 1.9A shows a sample of our NN. Seven components are the input of the neural net and the predicted score is the output of the neuron. We used this network to produce regression values. We made a 70%–30% split (i.e., 3500 images against 1500 images) of the dataset for training and testing, respectively. We trained our model with validation, first training the model with 3150 images and validating with 350 images. We used the trained model to predict the score of the 1500 test images and calculated our prediction score.

1.3.3 Dataset and user study

We used MIR Flicker [12] as our data source, and selected 5000 random images from all the available categories. Next, we conducted a user study to score selected images, with 374 participants in the survey, from whom we collected 12,748 scores on the 5000 images. The participants were given the following instruction: Rate the picture you are seeing, where 5 is the best score and 1 is the worst score.

Fig. 1.9B shows the histogram of user score against image count. At least 2 and a maximum of 3 users scored all the images to eliminate bias. We took the average of the user score as our target value.

1.4 Results and analysis

After applying our model to the training data, we have calculated prediction performance, which is listed in Table 1.1. A scatter plot of our predicted score against the user given score has been presented in Fig. 1.10A and its density has been plotted in Fig. 1.10B. The X axis of the scatter plot shows the user given score, the Y axis shows the predicted scores, and the points are the 1500 predicted scores. The color scheme of Fig. 1.10B represents the predicted score, where red is the high score and blue is the higher score. If we observe the plot, we can notice that most of our prediction lies in the range of the score 3 to 4.5. This is because the majority of user given scores are within this range. We can verify this by observing Fig. 1.10B. Also, we can observe that the lower score has a higher density from 3 on the plot, where higher score densities reach towards a score of 5. From this, we can conclude that images not having a

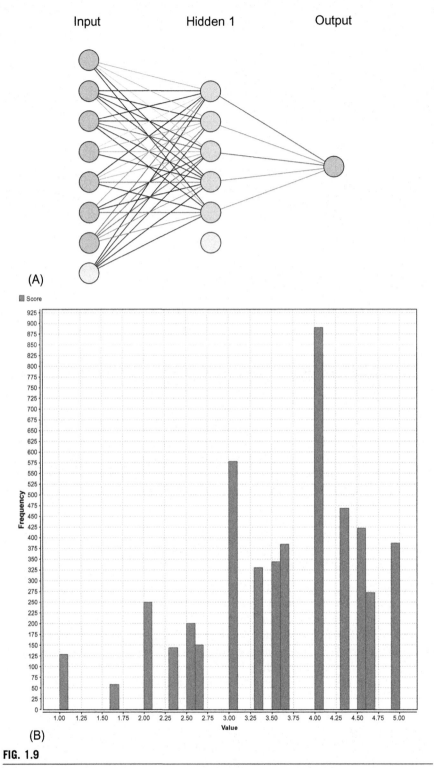

FIG. 1.9

Diagram of NN and dataset distribution. (A) Diagram of NN; (B) histogram of dataset.

Table 1.1 Performance score of NN and LR.

Attributes	Neural network	Linear regression
Root mean squared error	1.115 ± 0.196	0.940 ± 0.013
Absolute error	0.868 ± 0.170	0.761 ± 0.007
Relative error	$33.87\% \pm 6.84\%$	$28.37\% \pm 0.88\%$
Prediction average	3.615 ± 0.017	3.609 ± 0.013

higher score does not imply that those images do not have color harmony, but images with the higher score have a certain level of color harmony. This insight is similar to our result from previous work [13], where we showed that images with higher score maintain a certain level of rule of thirds, but the opposite is not true. This finding revalidates the findings of Mai et al. [14] in a comprehensive manner for the first time in computer vision literature, to the best of the authors' knowledge. Along with our NN results in Table 1.1, we have also listed results using a linear regression (LR) prediction model. Although LR seems to have a better performance, its result is more biased. Its predictions lie in the range of 3.50 to 3.74, as the majority of the dataset lies in this region, while root mean squared error (RMSE) is lower. Eq. (1.4) is the equation for LR.

$$x = -0.002 * \text{pca}_1 - 0.009 * \text{pca}_2 - 0.006 * \text{pca}_3 + 0.053 * \text{pca}_5 + 0.024 * \text{pca}_6$$
$$+ 0.027 * \text{pca}_7 + 3.615 \tag{1.4}$$

1.5 Discussion and future work

In this research, we have successfully developed a faster calculative method for extracting harmonic color features. We also have constructed a simplified method for perceiving beauty in images. The outcome of this research can be used not only in the domain of predicting aesthetics but also in other domains, i.e., cognitive model development, analysis of human perception of colors, color compositions, etc. We plan to create a model based on the human context of perceiving beauty as our future work. Along with this, we plan to combine features of this and our previous work [13] and create a robust model to perceive aesthetics with greater accuracy.

Acknowledgments

This research work is funded by North South University and we would like to thank all the participants from North South University who have participated in the user study of scoring the aesthetics of several images.

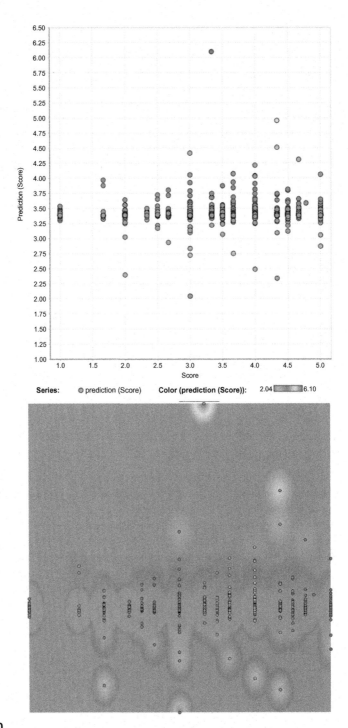

FIG. 1.10

Scatter and density plot of predicted scores. (A) Scatter plot; (B) density plot.

References

[1] P. Moon, D.E. Spencer, Geometric formulation of classical color harmony*, J. Opt. Soc. Am. 34 (1) (1944) 46.

[2] T. Stone, S. Adams, N. Morioka, Color Design Workbook: A Real World Guide to Using Color in Graphic Design, Rockport, 2017.

[3] P. Lu, X. Peng, X. Zhu, R. Li, An EL-LDA based general color harmony model for photo aesthetics assessment, Signal Process. 120 (2016) 731–745.

[4] P. Lu, X. Peng, C. Yuan, R. Li, X. Wang, Image color harmony modeling through neighbored co-occurrence colors, Neurocomputing 201 (2016) 82–91.

[5] R. Datta, D. Joshi, J. Li, J.Z. Wang, Studying aesthetics in photographic images using a computational approach, in: Computer Vision – ECCV 2006, Springer Berlin Heidelberg, 2006, , pp. 288–301.

[6] H. Phan, H. Fu, A. Chan, Color orchestra: ordering color palettes for interpolation and prediction, IEEE Trans. Vis. Comput. Graph. 24 (6) (2017) 1942–1955.

[7] C. Amati, N.J. Mitra, T. Weyrich, A study of image colourfulness, in: Proceedings of the Workshop on Computational Aesthetics—CAe'14, 2014.

[8] P. Lu, X. Peng, R. Li, X. Wang, Towards aesthetics of image: a Bayesian framework for color harmony modeling, Signal Process. Image Commun. 39 (2015) 487–498.

[9] M. Nishiyama, T. Okabe, I. Sato, Y. Sato, Aesthetic quality classification of photographs based on color harmony, in: CVPR 2011, 2011.

[10] Wikipedia. n.d. User:Datumizer—Wikimedia Commons. Retrieved from https:/commons.wikimedia.org/wiki/User:Datumizer.

[11] R. Keys, Cubic convolution interpolation for digital image processing, IEEE Trans. Acoust. Speech Signal Process. 29 (6) (1981) 1153–1160.

[12] M.J. Huiskes, M.S. Lew, The MIR Flickr retrieval evaluation, in: ACM International Conference on Multimedia Information Retrieval (MIR'08), Vancouver, Canada, 2008.

[13] A. Firoze, T. Osman, S. Psyche, R.M. Rahman, Scoring photographic rule of thirds in a large MIRFLICKR dataset: a showdown between machine perception and human perception of image aesthetics, in: 10th Asian Conference on Intelligent Information and Database Systems (ACIIDS), Lecture Notes in Computer Science (LNCS), Vietnam, 2018.

[14] L. Mai, H. Le, Y. Niu, F. Liu, Rule of thirds detection from photograph, in: Proceedings of the IEEE International Symposium on Multimedia (ISM), Dana Point, CA, USA, 2011, pp. 91–96.

Physiological parameter measurement using wearable sensors and cloud computing

2

X. Anitha Mary[a], Shoba Mohan[b], Suganthi Evangeline[c], K. Rajasekaran[d]

[a]*Instrumentation Engineering Department, Karunya Institute of Technology and Science, Coimbatore, India* [b]*Holoteq Group, Doha, Qatar* [c]*Electronics and Communication Engineering Department, Karunya Institute of Technology and Science, Coimbatore, India* [d]*Instrumentation Engineering Department, Karunya Institute of Technology and Science, Coimbatore, India*

2.1 Patient fall monitoring system

Elderly people are subject to frequent falls and this is the most significant cause of injury in this age group. Falls are the cause of many disabling fractures that could eventually lead to death or serious complications [1], such as infection or pneumonia. If proper treatment is not provided, serious health consequences may ensue [2]. An unexpected human fall can be detected by a 3-axis microelectromechanical system (MEMS) accelerometer. The human position and motion are detected using an accelerometer analog sensor. The longitude and latitude values from the accelerometer sensor detect the fall and the data are transmitted to the end user using wireless technology, such as a global positioning system (GPS) [3].

2.1.1 Methodology

The prototype shown in Fig. 2.1 consists of a set of biomedical sensors attached to the body of the person whose health condition is to be monitored. In this work, 3-axis accelerometers are used to detect a fall. These sensors are connected to an LPC2148 Advanced RISC Machine (ARM) microcontroller. The microcontroller receives the signals from the sensors and processes the data and checks for the condition of the person. If the condition is normal, then the microcontroller repeats the process of receiving the data from the sensors and monitoring the position of the person. Whenever the condition steps out of the normal range, the microcontroller checks for two or more values; if the same condition still remains, the microcontroller sends alert messages to the caretakers and/or concerned healthcare professionals about the unusual condition of the person being monitored. Once the doctor receives

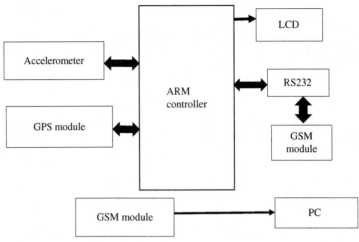

FIG. 2.1

Block diagram of fall detection system.

the message, an immediate response can be provided to the affected person. An SMS (short message service) is sent in response to the person in need of help through a global system for mobile communication (GSM) modem and the corresponding fallen location tracked by global positioning system (GPS) is displayed on the display unit connected to the controller via port pins.

2.1.2 **ARM LPC2148 microcontroller**

NXP Semiconductors (formerly Philips Semiconductors) designed a 32-bit microcontroller grouped into the LPC series. It has a CPU with emulation and debugging support and also with in-circuit programming features that support 32 kB to 512 kB high-speed flash memory. The microcontroller used in the prototype has a 128-bit wide memory interface with pipelined architecture, which enables faster code execution. An alternative instruction set for the ARM 32-bit is the Thumb 16-bit, which is ideal for critical code size applications [4].

Each microcontroller unit consists of the processor core, which is the heart of the core, and memory, including static RAM (SRAM) memory and flash memory. In some higher-level applications, cache memory is also used for storage purposes. The device also supports serial communications interfaces ranging from a universal serial bus (USB) 2.0 full-speed device, multiple universal asynchronous receiver and transmitter (UART), serial peripheral interface (SPI), and synchronous serial port (SSP) to an Inter-Integrated Circuit (I2C) and on-chip SRAM of 8 kB up to 40 kB. These added interfaces enable ARM-based products to handle applications in networking and also imaging very well. All ARM-based microcontrollers are provided with 32-bit timers, 10-bit ADC/DAC, and real-time clock (RTC). With the help of 45 fast GPIO lines with external interrupt pins, the ARM-based controller is suitable for industrial control and medical systems.

2.1.3 **ADXL 335 3-axis accelerometer**

ADXL335 is a 3-axis accelerometer provided with signal-conditioned voltage out-puts. It is known for compactness and low power consumption. It is used to measure acceleration with a full-scale range of $\pm 3\,g$. It can be used to measure both static and dynamic acceleration, such as tilt-sensing or acceleration resulting from motion, shock, or vibration [5]. It contains polysilicon surface-micromachined sensor and signal-conditioning circuitry, which enables implementation of open-loop accelera-tion measurement architecture. The changes of acceleration in three axes are mon-itored continuously and the output signals are directly proportional to acceleration. When the changes in acceleration exceed the threshold values, it is considered a fall. Fig. 2.2 shows the connection diagram of sensor interfacing with the LPC2148.

2.1.4 **Global positioning system**

The GPS provides location and time data regardless of climatic conditions. This is possible with the help of the signals it receives from an unobstructed line of sight from three or more GPS satellites. The system can be employed in defense, social, and commercial applications. The communication pattern used by the GPS is serial mode, in which the location of the person is sent to the concerned caretaker. The serial data are sent according to National Marine Electronics Association (NMEA) standards [6].

The first six bytes of data received from GPS module are decoded using an online decoder. The decoded output format is compared with the default GPS format. If both follows same format, say GPGLL, then the data is transmitted or else it will results in error. Once the format get matched, then the latitude and longitude values are extracted and displayed on an LCD and also transmitted to mobile phone as text message.

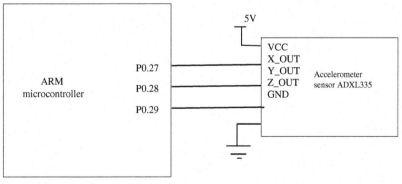

FIG. 2.2

Connection diagram of accelerometer sensor with ARM controller.

2.1.5 Global system for mobile communications

The European Telecommunications Standards Institute (ETSI) developed the GSM standard, which defines the protocols for second-generation (2G) digital cellular networks. Computers use AT commands to control modems. The merits of using AT commands are as follows: can able send short message from SIM300 module, can able to delete messages from inbox, can able to receive an incoming call, can able to dial to a new number using SIM300 module. GSM modems also support an extended set of AT commands.

2.1.6 Interfacing ADXL335 accelerometer sensor with LPC2148 microcontroller

The ARM7 LPC2148 uses three channels of ADC for reading X, Y, and Z axes from the ADXL335. The output of the accelerometer is sent to the ADC channel resulting in a digitized form of 10-bit representation. The digital output of ADC is converted into voltage by the formula:

$$\text{Output (V)} = \text{Output (digital)} * \frac{3.3}{1023} \tag{2.1}$$

The values of the coordinates are displayed on LCD. The same data is transmitted to the mobile phones of the caretaker/relatives of the fallen person, using GSM as SMS.

2.1.7 Calibration

The LPC2148 has a 10-bit ADC:

$$\text{Step size} = \frac{V_{ref}}{2^{10}} \tag{2.2}$$

where

V_{ref} —used to detect step size

$$D_{OUT} = \frac{V_{input}}{\text{Step size}} \tag{2.3}$$

where

D_{OUT} —decimal output digital data.
V_{input}—analog input voltage.

Table 2.1 summarizes the zero bias values and the equivalent voltage levels given to the analog-to-digital converting unit and Table 2.2 summarizes the calculated digital value for various *g* ranges. The experiment of interfacing a triple-axis accelerometer with an ARM microcontroller is conducted by varying the axis position in order to find the threshold value. The threshold values for different types of falls have been recorded and tabulated in Table 2.3.

Table 2.1 Zero bias level of X, Y, and Z axes.

Zero *g* bias level (ratiometric)	Conditions	Min	Typ	Max	Unit
0*g* Voltage at X_{OUT}, Y_{OUT}	$V_S = 3V$	1.35	1.5	1.65	V
0*g* Voltage at Z_{OUT}	$V_S = 3V$	1.2	1.5	1.8	V
0*g* Offset vs. Temperature			±1		mg/°C

Table 2.2 Calibration of ADXL335.

	Range	Voltage	ADC value
ADXL335	0g	1.35	1A3
	1.2g	1.62	1F6
	1.5g	2.02	273
	1.8g	2.43	2F2
	2g	2.70	346
	2.2g	2.97	399
	2.4g	3.04	3ED

Table 2.3 Threshold value for ADXL335.

Position	X-AX-X-AXISXIS	Y-Y-AXIS	Z-Z-AXISAXIS
Normal	0.319	1.274	1.650
Forward fall	0.322	1.687	1.970
Backward fall	0.319	1.487	1.380
Left side fall	0.312	1.690	1.650
Right side fall	0.325	1.512	1.85

By comparing all the values, we have set the threshold values as: X=0.322V, Y=1.320V, and Z=1.700V.

2.1.8 Interfacing GPS and GSM with LPC2148

Fig. 2.3 shows how to interface the GPS and GSM to the LPC2148 microcontroller. The GSM and GPS are connected to the ARM7 LPC2148 through UART0. In the place of COM2 in Fig. 2.3, either GPS or GSM can be connected. A text message may be sent through the modem by interfacing only three signals: TX, RX, and GND. The GPS connection requires only TX and GND. The transmitter of the GPS is connected to the receiver of UART0. The GPS transmits the latitude and longitude and the data is received.

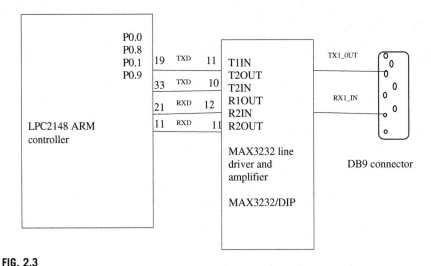

FIG. 2.3

Circuit diagram of interface of GPS and GSM with LPC2148.

2.2 Monitoring of patient body temperature

The most predominant symptom of an infection is a rise in temperature. Temperature is also one of the main indications used to detect symptoms related to stress, which ultimately leads to severe health conditions like stroke or heart attack. The measurement of body temperature is extremely useful for determining physiological conditions as well as for other things, such as activity classification or even harvesting energy from body heat [7]. The main aim of this work is to monitor the temperature of the human body over a period of time and update it in the cloud. This system will enable physicians to be virtually connected with their patient and to use the temperature data in arriving at a diagnostic conclusion. For the patient, it is a wearable sensor that can be used anywhere and anytime [8].

2.2.1 Methodology

In order to measure the body temperature of an individual, the wearable device is connected to the person's wrist or any other comfortable body part where the thermistor is in contact with the body such that the temperature can be monitored. The data is then sent to the microcontroller and from the microcontroller to the server through the Internet of Things (IoT). The temperature is monitored on the website, as shown in Fig. 2.4. The doctor has access to the patient's temperature and ensures that any requirements are given according to the schedule [9].

The wearable temperature sensor module components include a thermistor, Arduino Pro Mini microcontroller board, and the ESP8266 Wi-Fi module. The Wi-Fi module is connected to the Internet. When the thermistor detects the temperature, the data are sent to the ThingSpeak application and it saves the value for future use.

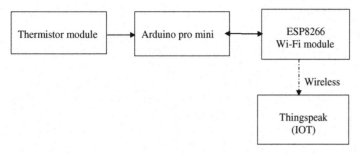

FIG. 2.4

Block diagram of patient body temperature measurement system.

2.2.2 ESP8266 Wi-Fi module

The ESP8266 is a low-cost Wi-Fi module with full TCP/IP stack and CPU, which allows TCP/IP communication between microcontrollers to a Wi-Fi network using Hayes-style commands. The flash memory capacity is 1 Mb. It is low cost and can be useful for many Wi-Fi applications [10].

2.2.3 ATmega328 microcontroller

The Arduino microcontroller board is based on the ATmega328 architecture. It has a Harvard architecture with memory capability of 32 K flash memory, 2 K SRAM data memory, and 1024 bytes of EEPROM. Additional features include 32 8-bit general-purpose registers, 3 timer/counters, 6-channel 10-bit ADC, 14 digital input/output pins, and 6 analog inputs. The operating voltage is between 1.8 and 5.5 V [10].

2.2.4 LM393 thermistor module

The thermistor module is used to detect the temperature change, working as a dual differential comparator. Its advantages include low operating voltage, low power consumption, and compactness, the primary specifications in circuit design for portable consumer products [11].

2.2.5 Software module ThingSpeak

ThingSpeak is an open source IOT application using HTTP protocol over the Internet, mainly used to update the measured sensor values at regular intervals. Its features include the ability to collect and share data from private channels. It involves MATLAB support for analysis and visualizations. It also responsible for creating alerts, event scheduling, and apps integration, and it has wide community support [12].

2.2.6 **Experimental setup**

The ESP8266 Wi-Fi module is configured using AT commands by setting to the preferred mode for communication. The thermistor senses the temperature and sends the data to the microcontroller, which is responsible for converting the data into three different scales, i.e., Kelvin, Fahrenheit, and Celsius. The sensed temperature values are uploaded in the Thingspeak cloud. Once the ESP8266 module get connected with internet, the sensor values are sent to cloud. In order to access the data available in the cloud, the read application program interface (API) keys should be used in the programming module. The doctor or care giver can regularly analyse the temperature values using the cloud services. The ESP8266 Wi-Fi module is configured in station mode so that the module is getting connected with internet and it is used to upload the sensor values in the cloud. The internet provider username and password are given to the Wi-Fi module in order to have a connection with internet.

2.2.7 **Results and discussion**

The wearable temperature monitoring experimental setup has been verified. As mentioned earlier, the results are sent to the cloud, where concerned persons can access them anywhere using an intranet and the Internet. Sample temperature values expressed in Celsius through ThingSpeak are shown in Fig. 2.5.

2.3 **Vital parameter measurement using PPG signal**
2.3.1 **Significance of remote patient monitoring**

Remote patient monitoring (RPM) plays a significant role in monitoring of patients, especially the elderly. Physiological parameters such as glucose level, oxygen saturation level, hemoglobin, and blood pressure can be continuously monitored on the patient end. Through wireless technology, these parameters can be transmitted to the physician or other medical staff, where immediate action can be taken. This allows better monitoring of patients, especially postsurgery. Remote monitoring systems have an added advantage over the conventional method of wire connection between the patient and doctor. The literature study reveals a method of wireless communication through radio frequency to transmit a PPG signal.

2.3.2 **Importance of PPG signal**

The word plethysmograph is a combination of two ancient Greek words *plethysmos* meaning increase [13] and *graph*, the word for write [14]. A photoplethysmograph (PPG) uses an optical signal to measure blood volume changes [15]. A light source and a sensor-fitted probe are used to detect the cardio-vascular pulse wave from the body. The blood movement in the vessels, going directly from the heart to the fingertips and toes, is acquired by the PPG device. The PPG signal, which is a wavelike

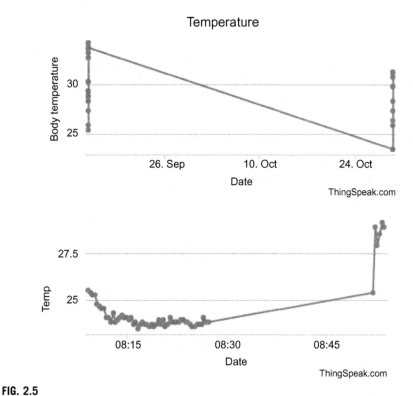

FIG. 2.5

Sample temperature value in ThingSpeak.

motion, reflects this blood movement. This optical measurement technique uses an invisible infrared light sent into the tissue and the amount of backscattered light corresponds with the variation in blood volume [14]. Hertzman was the first to find a relationship between the intensity of backscattered light and blood volume in 1938. The low cost and simplicity of this optical-based technology offers significant benefits to healthcare (e.g., in primary care where noninvasive, accurate, and simple-to-use diagnostic techniques are desirable). Further development of PPG could place this methodology among other important tools used in the management of vascular disease [16].

2.3.3 Experimental setup

The PPG signal is obtained using a PPG sensor and the obtained PPG data are processed using LabVIEW software from National Instruments; the actual heart rate is measured from the PPG signal using a threshold peak detector. Any abnormal heart rate such as bradycardia or tachycardia can be detected. The result is obtained from

LabVIEW by an alarm set for any abnormal heart rate signal. The patient's data from the PPG is transmitted and received through a Bluetooth module. Using LabVIEW provides a visual view of the heart rate data. With LabVIEW, the patient himself or care-provider can able to visualize the performance of his heart rate. The PPG data available to doctor as well as to his patient also. This is made possible using the LabVIEW environment.

2.4 Oxygen saturation measurement using PPG signal

Oxygen saturation level measurement can be monitored using a pulse oximeter. The pulse oximeter determines arterial oxygen saturation of the blood (SaO_2) and it is denoted as SpO_2. The pulse oximeter consists of red and IR LEDs and a photodiode, and can be classified into reflectance or transmittance types.

The red (660 nm) and the IR (940 nm) PPG signals are detected by the photodiode, and its output is converted to a voltage by a transimpedance amplifying circuit and is given to the data acquisition system after appropriate amplification. The acquired analog signals are taken into the LabVIEW platform, which hosts SpO_2 on a realtime basis. The photodetector produces a current signal proportional to the light intensity transmitted through the finger from LEDs. The current signal produced by the photodiode is converted to a voltage signal at the output node of the transimpedance amplifier. Photodiodes are used for light detection, conversion, and measurement.

2.4.1 SpO$_2$ calculation

The signals are first fed to the SpO_2 program, which calculates the saturation of oxygen in the blood. First the maximum and minimum peaks of the PPG signal are found and, using these values, SpO_2 is calculated. The normal range of a PPG signal is 0.5–5 Hz. The high-frequency noise can be removed by using a low-pass filter with a cut-off frequency of 5 Hz. For calculation of SpO_2, the V_{max} and V_{min} components of both the red and IR signals are extracted.

2.5 Heart rate measurement using ECG signal
2.5.1 Monitoring of ECG signal over wireless transmission

Fig. 2.6 shows a block diagram of heart rate measurement using an ECG signal. The ECG from the patient is obtained using three-lead, ring type electrodes. Electrode gel is used for obtaining better contact between the electrode and the patient. The LabVIEW program detects heart rate from the ECG and displays whether the patient is normal or abnormal. The obtained heart rate is sent to the physician's mobile unit using a Bluetooth module. The conditions for heart arrhythmia are that, if the heart rate is above 100 bpm then it is tachycardia, and if the heart rate is less than 60 bpm it is bradycardia; if the heart rate is above 150 bpm it is atrial flutter. Fig. 2.6 shows the

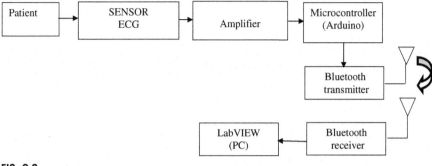

FIG. 2.6

Experimental setup for heart rate measurement system.

experimental setup for this ECG heart rate measurement system. Using LabVIEW, the ECG signal is acquired and processed by placing ring electrodes on chest and it is monitored continuously to detect any abnormal alterations in the heart rate at an early stage.

2.5.2 Results and discussion

The real-time signal is acquired from the patient using the electrodes and it is implemented using LabVIEW. The acquired signal is analyzed to distinguish whether the obtained signal is a normal signal or a tachycardic or bradycardic cardiac signal. If the signal obtained is an abnormal signal, then an alarm is set to indicate a patient with abnormal ECG. The ECG heart rate measurement uses baseline drift estimation to eliminate noise from the signal. The heart rate is detected using the R-R peaks from the baseline drift elimination signal.

The real-time signal is read in LabVIEW, the signal is processed, heart rate is detected, and any abnormality in the heart rate is obtained. If an abnormal heart rate is detected, an alarm is shown on the front panel of LabVIEW. The filtered ECG signal is displayed, and the output of the signal, either normal or abnormal, is displayed using LEDs in LabVIEW.

The significance of wireless heart rate detection is, the information is transmitted to any end devices which the user want to view. That is, the end device can be a mobile or PC, using Bluetooth technology the data can be transmitted.

We have built a prototype using parts of the case from a commercial blood pressure monitor for the wrist. All necessary sensors are placed inside (acceleration and pressure) or on the outside (temperature) of the case, except for the ECG electrodes, which are placed on the chest and connected by cables. The device is powered by the system itself. It currently transmits every sample of the sensor raw data wirelessly to the stationary computer, where the data is stored and processed with a common time base. Our prototype only collects and forwards the data and does not yet perform any calculations for data compression or analysis.

2.6 Conclusion

With wireless technology, these physiological parameters can be monitored continuously and warning and causality services can be provided to the medical staff and doctors, as early as possible. Patient fall detection and body temperature measurement use wireless technology. The remote monitoring of the PPG signal is also useful in continuous monitoring of vital parameters, especially for the postoperative patient. The doctor can view the patient details at the remote end. Heart rate and SpO_2 parameters can be transmitted wirelessly using the PPG signal, and heart rate is also measured using an ECG signal.

References

[1] Nihseniorhealth, About falls, Available online: http://nihseniorhealth.gov/falls/aboutfalls/01.html, 2013. Accessed 10 December 2013.

[2] F.W. Booth, C.K. Roberts, M.J. Laye, Lack of exercise is a major cause of chronic diseases, Compr. Physiol. 2 (2) (2012) 1143–1211 Author manuscript; available in PMC 2014 Nov 23.

[3] S.R. Lord, H.B. Menz, C. Sherrington, Home environment risk factors for falls in older people and the efficacy of home modifications, Age Ageing 35 (Suppl 2) (2006) ii55–ii59.

[4] LPC2148, Microcontroller, August, https://www.nxp.com/docs/en/data-sheet/LPC2141_42_44_46_48.pdf, 2011.

[5] Sparkfun, Analog devices Small, Low Power, 3-Axis ±3g Accelerometer, https://www.sparkfun.com/datasheets/Components/SMD/adxl335.pdf, 2009.

[6] Introduction to GPS—US Environmental protection agency, https://www.epa.gov/sites/production/files/2015-10/documents/global_positioning_system110_af.r4.pdf, 2015.

[7] M. Aminian, H.R. Naji, A hospital healthcare monitoring system using wireless sensor networks sensor networks, J. Health. Med. Inform. 4 (2013) 121, https://doi.org/10.4172/2157-7420.1000121. Int. J. Adv. Netw. Appl. 4 (1) 1497-1500 (2012). ISSN: 0975-0290.

[8] S.C. Mukhopadhyay, Wearable sensors for human activity monitoring, IEEE Sens. J. 15 (3) (2015) 1321–1330.

[9] S. Aruna Devi, S. Godfrey Winster, S. Sasikumar, Patient health monitoring system (PHMS) using IoT devices, Int. J. Comput. Sci. Eng. Technol. 7 (03) (2016) 68–73.

[10] Atmel, 8-bit AVR Microcontrollers, http://www.atmel.com/Images/Atmel-42735-8-bit-AVR-Microcontroller-ATmega328-328P_Datasheet.pdf ESP8266EX Datasheet 2015, 2016. https://cdn-shop.adafruit.com/product-files/2471/0A-ESP8266__Datasheet__EN_v4.3.pdf.

[11] Onsemi, LM393, LM393E, LM293, LM2903, LM2903E, LM2903V, NCV2903 On semiconductor, https://www.onsemi.com/pub/Collateral/LM393-D.PDF, 2016.

[12] D. Kyriazis, et al., Sustainable smart city IoT applications: heat and electricity management & eco-conscious cruise control for public transportation, in: 2013 IEEE 14th International Symposium and Workshops on World of Wireless, Mobile and Multimedia Networks (WoWMoM), 2013, 2013, pp. 1–5.

[13] K. Shelley, Photoplethysmography: beyond the calculation of arterial oxygen saturation and heart rate, Anesth. Analg. 105 (2007) S31–S36.

[14] M. Alnaeb, N. Alobaid, A. Seifalian, D. Mikhailidis, G. Hamilton, Optical techniques in the assessment of peripheral arterial disease, Curr. Vasc. Pharmacol. 5 (2007) 53–59.

[15] S. Banka, M.X. Anitha, Remote monitoring of heart rate, blood pressure and temperature of a person, Int. J. Emerg. Technol. Comput. Sci. Electron. 8 (1) (2014).

[16] M. Elgendi, On the analysis of fingertip photoplethysmogram signals, Curr Cardiol Rev. 8 (1) (2012) 14–25. Review.

Further reading

Addison, J. Watson, A novel time–frequency-based 3D Lissajous figure method and its application to the determination of oxygen saturation from the photoplethysmogram, Meas. Sci. Technol. 15 (11) (2004) L15.

Ashisha, A. Mary, Rajasekaran, Jegan, IoT-based continuous bedside monitoring systems, Adv. Intell. Syst. Comput. 750 (2019) 401–410.

P. Cheang, P. Smith, An overview of non-contact photoplethysmography, in: Electronic Systems and Control Division Research, Department of Electronic and Electrical Engineering, Loughborough University, UK, 2003, pp. 57–59.

Global System for Mobile Communication (GSM), Requirements for GSM operation on railways, European Standard (Telecommunications Series), 2001. http://www.etsi.org/deliver/etsi_en/301500_301599/301515/01.00.00_20/en_301515v010000c.pdf.

M. Johnson, R. Jegan, X. Anitha Mary, Performance measures on blood pressure and heart rate measurement from PPG signal for biomedical applications, in: Proceedings of IEEE International Conference on Innovations in Electrical, Electronics, Instrumentation and Media Technology, ICIEEIMT 2017, January, 21 November 2017, vol. 2017, 2017, pp. 311–315.

K.S. Pavithra, A. Mary, Rajasekaran, Jegan, Low cost non-invasive medical device for measuring hemoglobin, in: Proceedings of IEEE International Conference on Innovations in Electrical, Electronics, Instrumentation and Media Technology, ICIEEIMT 2017, January, 21 November 2017, vol. 2017, 2017, pp. 197–200.

Rajasekaran, S.J. jyothi, S. Rekh, A. Mary, Non-invasive hemoglobin measurement: a great blessing to the rural community, md Curr. J. (2015).

K. Rajasekaran, D.X.A. Mary, R. Jegan, Smart technologies for non-invasive biomedical sensors to measure physiological parameters, in: Handbook of Research on Healthcare Administration and Management, IGI Global Publication, 2016.

K. Rajasekaran, X.A. Mary, R. Jegan, Smart technologies for non-invasive biomedical sensors to measure physiological parameters, in: Handbook of Research on Healthcare Administration and Management, IGI Global Publication, 2016.

M. Sindhu, X. Anitha Mary, Development of real-time, embedded data monitoring wireless networking system to characterized solar panel, Pak. J. Biotechnol. 13 (2016) 265–269.

T. Ubbink, Toe blood pressure measurements in patients suspected of leg ischaemia: a new laser doppler device compared with photoplethysmography, Eur. J. Vasc. Endovasc. Surg. 27 (2004) 629–634.

N. Unno, K. Inuzuka, H. Mitsuoka, K. Ishimaru, D. Sagara, Automated bedside measurement of penile blood flow using pulse-volume plethysmography, Surg. Today 36 (3) (2006) 257–261.

Social media data analytics using feature engineering

J. Anitha[a], I-Hsien Ting[b], S. Akila Agnes[a], S. Immanuel Alex Pandian[c], R.V. Belfin[a]

[a]*Department of Computer Science and Engineering, Karunya Institute of Technology and Sciences, Coimbatore, India*
[b]*Department of Information Management, National University of Kaohsiung, Kaohsiung, Taiwan*
[c]*Department of Electronic and Communication Engineering, Karunya Institute of Technology and Sciences, Coimbatore, India*

3.1 Introduction

The rapid growth of the social network has exponentially increased the volume of social media data on the Internet. These data are generated by public users through their profile statuses, posts, comments, tweets, and reviews that carry the message in the form of image, text, and video. Social media like Twitter has created a platform for people to share and expose their views on social and political issues, trends, natural crises, and other kinds of information. The user-generated data may cover high-quality and valuable information, but since the data is received from diverse sources, it is fuzzy and unstructured in nature. Hence an efficient technique is required to extract and analyze the essential information from the huge amount of data generated. Data analytics plays an important role in the field of big data, focusing on understanding the encapsulated information in the huge heterogeneous unstructured data.

Different techniques have been developed to understand trending images and content in social media. People share trending images through social networks by tagging related comments. Identifying similar media images or contents from the huge amount of shared data is a challenging task and this has become increasingly popular in social media. In the last decades, several machine-learning approaches have been proposed to improve the performance of social data analytics. These approaches extract multimodel feature vectors [1] from social web data and these feature vectors are used to label the social images.

A huge amount of data is generated via social media in unstructured format websites [2, 3]. The ideas shared in the social networks, such as Facebook and Twitter, help to analyze people's opinions about recent trends and social issues. The evolution of big data in social media has brought more importance to the field of social data analytics. Recently many research projects have been proposed to analyze the facts

Systems Simulation and Modeling for Cloud Computing and Big Data Applications
https://doi.org/10.1016/B978-0-12-819779-0.00003-4

behind the shared data using various data-mining and machine-learning techniques. Sentiment analysis for mining opinions, sentiments, and emotions has been reported [4], which can assist in making decisions in several situations such as crisis and business management. Gandomi and Haider [5] have presented various analytics techniques to mine intelligence from unstructured multimedia data [5].

Extracting intelligence from social images influences the development of effective social data analytics techniques in the area of image recognition and information retrieval systems [6]. Manual annotation of social images is a very expensive and time-consuming process. Deepwalk [7] is an approach that uses local information to encode the social relationship using unsupervised feature learning.

Over the last decade, research in social media has become popular and many machine-learning algorithms have been developed to transform unstructured social media data into representation vectors. Various techniques [8] for handling the challenges in data gathering, data preparation, and topic discovery in social media analytics have been proposed. However, efficient feature engineering techniques to represent the social media data in various mining applications are needed.

This chapter discusses various feature engineering approaches for representing and categorizing heterogeneous unstructured data. The rest of the chapter is organized as follows: Section 3.2 describes the proposed framework for social media data analytics; Section 3.3 explains the recent feature engineering techniques used for extracting features from the text, image, and social relationships; clustering methods in big data analytics are reported in Section 3.4; and, finally, Section 3.5 presents the conclusion.

3.2 Social media data analytics

A data analytics framework is proposed to encode unstructured social media data into a meaningful feature vector and cluster it with respect to the target application. The social images can be collected from Flickr[a] sources and metadata of each picture such as title, description, location, view count, upload date, uploader's name, tags, comment threads of the picture, etc. can be obtained through Flickr API.[b] Based on the tags, comments, and network relationship details, the proposed model labels the images using clustering techniques. Image labeling is a technique that provides contextual information about the image. It is required in many big data applications to analyze the views and responses of people about current trend and issues. The proposed framework labels images based on its visual content and information provided by the public. The proposed framework for social media data analytics is shown in Fig. 3.1. Since the input contains a variety of data (text, image, relationship graph), this needs to be vectorized using machine-learning and soft-computing techniques. Finally, the vector data are grouped into clusters using clustering techniques. The grouping of such images that originate from social media helps in many applications to analyze the data.

---

[a]https:/snap.stanford.edu/data/web-flickr.html.
[b]https:/www.flickr.com/photos/tags/dataset/.

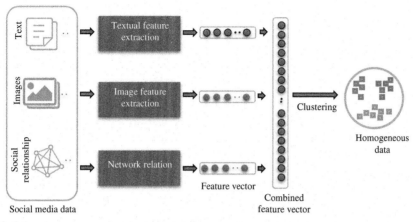

FIG. 3.1

Proposed framework for social media data analytics.

3.2.1 **Social media data format**

People freely share their data in social media, which normally includes images, comments, and tags. In addition to that, the social images also have network relation information and all these details are useful for efficient content analysis. The network details are built by forming edges between the images shared by the same user, commented by the same user, and annotated with the same tags. Fig. 3.2 shows the characteristics of

FIG. 3.2

Social media data characteristics.

social media data. Unfortunately, these social media data are not in a structured format and need to be transformed into a structured format for further data analysis.

3.3 Feature representation

The real-world social media data are represented in various formats like text, images, videos, etc. This data cannot be used directly for analysis. Feature engineering is a process of extracting feature vectors from the raw data and these vectors are further used in classification and clustering applications. Features can be extracted in two ways: traditional extraction and representation learning approaches. The proposed data analytics framework includes three characteristics of social media data—text, image, and social relation—for analysis. Earlier research [1, 9] confirms that the combination of visual, textual, and semantic representations improves the efficiency of the analysis technique.

3.3.1 Textual feature extraction

In social media, tags and comments are raw text that comprises a confined number of characters that regulate the flexible order of text, such as spaces, line breaks, or tabulation. The raw text version is different from the formatted version, where style information is combined. From the structured text, structural components of the document such as paragraphs, subdivisions, and related elements are identified.

In analyzing the textual data, the raw text must to be preprocessed by removing the unwanted elements from the document. Preprocessing is the process of tidying and arranging the text for its analysis using preprocessing procedures. The existing preprocessing methods are handling expressive lengthening, emoticon handling, HTML tag removal, punctuation handling, slang handling, stop word removal (such as "a," "the"), stemming, and lemmatization (the last two are covered in Section 3.3.1.3) [10]. The text analytics preprocessing steps are shown in Fig. 3.3.

3.3.1.1 Tokenization

Tokenization [11–14] is a step that breaks longer strings of text into shorter bits, or tokens. The tokens are also called shingles or n-grams. n-grams is the sequence of n words in the sentence. Extensive chunks of text are tokenized into sentences, sentences tokenized into words, etc. More processing is generally performed after a piece of a paragraph has been appropriately tokenized. Tokenization is also known as text segmentation or lexical analysis. Sometimes segmentation is utilized to refer to the division of a massive chunk of text into pieces longer than words (e.g., paragraphs or sentences), while tokenization is reserved for the division process that results exclusively in words. There are several tokenizers available as open source and for commercial use. The open source package for tokenization in R is given in Mullen et al. [11]. The classification of the tokenization process with respect to the number of words is shown in Fig. 3.4.

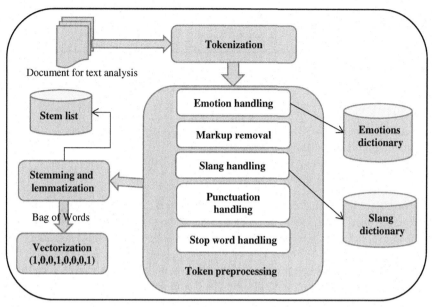

FIG. 3.3

Preprocessing steps in text mining.

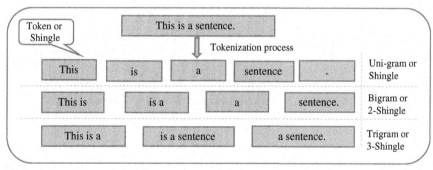

FIG. 3.4

Tokens and n-grams.

3.3.1.2 Token preprocessing

Removal and replacement of unwanted tokens is an essential part of the text preprocessing framework. There are many other significant steps to be followed in the token preprocessing. They include,

- Convert all text to lower case.
- Remove numeric values (if they are out of context).
- Remove punctuations and white spaces.

FIG. 3.5

Stop word removal.

- Remove stop words.
- Handle markups, such as <html>.
- Replace different slang tokens with general English slang.
- Replace emoticon symbols with related text.

Stop words are words that do not contribute much to the overall meaning of the sentence, e.g., "the," "and," and "a." An example of the stop word removal process is given in Fig. 3.5. The words in the sentence that are struck through represent the stop words.

It is clear that the token preprocessing relies on prebuilt dictionaries, databases, and rules. As shown in Fig. 3.3, the stem list, slang database, and the emotions dictionary are prebuilt dictionaries and databases.

3.3.1.3 Stemming and lemmatization

Stemming [10] and lemmatizing [15, 16] are usually acknowledged as sibling methods and placed under the same roof [17]. Both have the similar purpose of reducing the variant words in the input text. The prime contrast between these processes lies in their outputs. The output result of stemming is a stem, and that of lemmatization is a lemma.

The stem is the part of a word applied to form new words through linguistic techniques such as compounding (e.g., six-pack, daydream) or affixation (e.g., perishable, durable). The stem may be a valid, stand-alone word (free stem) or a partial word that requires an affix to make a word (bound stem). In the preceding example words, *perish* is a free stem and *dur* is a bound stem. A classification of stemming techniques is provided in Fig. 3.6.

Lemmas, on the other hand, are stand-alone linguistic elements and a vocabulary form of a lexeme. A lexeme is the set of all the alternative forms of a word that have the same meaning, and a lemma is the one base word that represents the lexeme. For example, *sit, sat, sits, sitting* is a lexeme represented by the lemma *sit*. A huge amount of research has been applied to stemming and lemmatization in the field of text analytics. A simple lemmatization example is shown in Fig. 3.7.

The lemmatization process is followed by the calculation of the frequency of words. A simplified method for representing documents is the bag-of-words model (BoW). This representation method only incorporates the words and their corresponding recurrences in the text, independent of their positions in the sentence or document. The bag of words for the example is shown in Fig. 3.8.

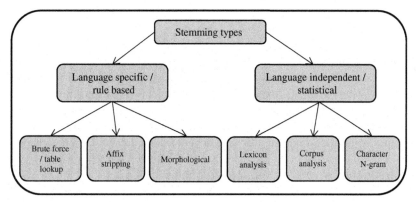

FIG. 3.6

Stemming classification.

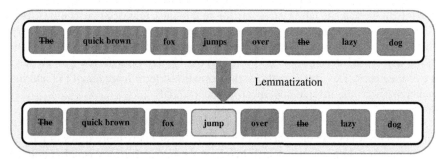

FIG. 3.7

Lemmatization process.

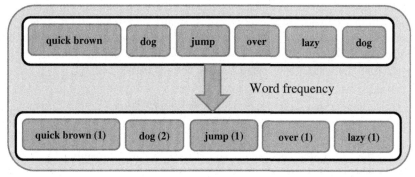

FIG. 3.8

Bag of words.

3.3.1.4 Vectorization

Vectorization [15, 18–22] is the process of defining an algebraic model to describe the text documents as vectors of tokens or index terms. Each dimension in the model represents a different term. Typically terms are single words, keywords, or larger phrases. If words are picked to be the terms, the dimensionality of the vector is the count of words in the dictionary (the sum of distinct words appearing in the corpus). The weight for all the terms occurring in the document is a nonzero integer. Several methods are available to compute these weights. One of the best possible techniques is the term frequency-inverse document frequency (TF-IDF) method.

Term frequency-inverse document frequency

Usually, TF-IDF values are utilized in knowledge mining and text analytics. This value is a statistical metric adopted to assess the significance of a word in a corpus. The importance grows proportionally to the count of times a word appears in the corpus. However, this is also balanced by the incidence of the word in the corpus. Search engines frequently adopt a modification of the TF-IDF weighting method as a principal tool in scoring and rating a document's significance related to a user query. One of the simple rating functions is estimated by summing the TF-IDF for every query term. The subject areas including text summarization and classification use TF-IDF for the stop-word filtering process. Typically, the TF-IDF is measured with the help of two measures. The initial metric is the normalized term frequency (TF) and the second metric is the inverse document frequency (IDF).

- **TF:** Term frequency is a measure that calculates how often a word occurs in a document. It is probable that a word would appear many more times in long documents than in shorter ones. Therefore, the term frequency is defined in Eq. (3.1) as

$$TF(t,d) = \frac{Count\ of\ term\ t\ in\ the\ document\ d}{Count\ of\ terms\ in\ the\ document\ d} \tag{3.1}$$

- **IDF:** Inverse document frequency is a measure that estimates the importance of the term. In calculating TF, every word is equally significant. But it is clear that some words, such as *is*, *of*, and *that*, may occur many times yet have limited importance. Therefore the idea is to lower the value of common terms while scaling up the unique ones using IDF, as defined in Eq. (3.2):

$$IDF(t,D) = \log_e \frac{Count\ of\ documents}{Count\ of\ documents\ with\ term\ t} \tag{3.2}$$

where t is the term in the document d and D is the set of documents $\{d \in D\}$. The possible weights for the vectors can be a binary term vector or a weighted term vector. For a binary term vector, the weights are denoted as 1 for the term occurrence and 0 for the nonoccurrence of the term. Similarly, for the weighted term vector, the

Table 3.1 Binary and weighted term vectors.

	Possible terms	The, Quick, brown, dog, jump, over, lazy	
#	Document term	Binary term vector	Weighted term vector
1	the quick brown dog	(1,1,1,0,0,0)	(0.01, 1,2,0,0,0)
2	jump over lazy	(0,0,0,1,1,1)	(0,0,0,1,0.5,1)

relatively important terms will have more weight value. Examples of binary and weighted term vectors are given in Table 3.1.

3.3.2 Image feature extraction

Image feature extraction is the process of converting 2D image data into a set of numerical values, which are called as feature vectors. These feature vectors are used to describe the higher-level information of an image with a reduced dimension. Analyzing the visual content from the enormous amount of social image data is a complicated and time-consuming task. The complexity of the task can be reduced by transforming the vague high-dimensional image data into meaningful low-dimensional feature data. Since the social images are attained from different sources, the tool used for describing the image should be robust to image size and orientation. The existing feature extraction techniques are broadly classified into two categories: engineering methods and representation learning methods. The categorization of visual feature extraction techniques is presented in Fig. 3.9.

The traditional engineering approach uses handcrafted features that require domain level expertise. Based on the working location, the engineering techniques are further classified into two types: global methods and local methods. The global methods, including color histograms and histograms of oriented gradients (HOGs), extract the representative features from the entire image properties. Global feature descriptors represent the abstract content of an image; therefore this representation is computationally efficient. As global features are not influenced by image noise, these features are appropriate for raw data analytics. Local features are the standard

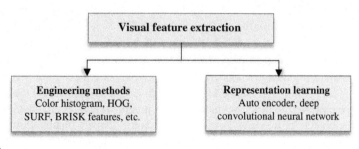

FIG. 3.9

Categorization of visual feature extraction techniques.

method to describe the images. The local feature extraction approaches extract features from a small portion of an image. Local features are focused on interesting points and not affected by occlusion, viewpoint, and illumination changes. Local image descriptors such as SIFT (scale-invariant feature transform) [23] and SURF (speeded-up robust feature) [24] are most powerful in classifying the objects, irrespective of image shape and orientation. SURF detectors identify the objects regardless of the image scale invariance. A few researchers [25] have suggested the fusion of local and global features that are used as image descriptors for content-based image retrieval.

Representation learning allows the model to learn various levels of features from images directly with respect to the target application. Deep convolutional neural network (DCNN) models can learn features from a low level, such as corners and edges, to high level. Representation learning can be done in two ways: (1) supervised feature learning, and (2) unsupervised feature learning. The deep CNN model is a supervised technique that can be used as a feature extractor by fetching the features from the last convolutional layer of the network. An autoencoder is an unsupervised learning technique to learn the features from the image through a series of encoding and decoding operations. Autoencoder-based feature extraction techniques have overcome the traditional feature extraction methods by adopting the unsupervised learning pattern.

3.3.2.1 Color histogram features

Color plays an important role in analyzing the characteristics of an image. A color histogram [26] describes the color-based features of an image that provide the probability distribution of the intensity levels in the image. The color histogram probability $hist(i)$ at intensity level i is calculated using Eq. (3.3).

$$hist(i) = \frac{Number\ of\ pixel\ at\ intensity\ level\ i}{Number\ of\ pixels\ in\ the\ image} \tag{3.3}$$

The color histogram can be obtained for various color models such as RGB, HSV, L/a/b/, YCbCr, etc. The color model describes the standard way to create colors from the primary colors of the color model. The RGB color model defines the pixel color by combining the intensity of red, green, and blue. The color features using the color difference histogram (CDH) have improved the performance of image retrieval as compared with other feature-based approaches [27].

3.3.2.2 Histogram of oriented gradients features

The histogram of oriented gradients (HOG) is a global feature extractor that extracts the shape faces of an image. Shape features are invariant to geometrical transformation and are preferred to describe the objects in image classification. HOG features describe the objects by dense and overlapping features that are obtained from intensive image scanning [28]. HOG features enhance the efficiency of the classification system by identifying the precise features of an image. HOG samples an image into small blocks using a kernel. It discovers the object shape using the gradient information of the pixels within the block. The image gradient determines the direction of the pixel from the intensity change between its neighboring pixels.

3.3.2.3 SURF features

The SURF algorithm is used to detect invariant scale blob features. Blob detectors obtain the subregions of an image in which all pixels have similar properties as compared to neighboring regions. SURF uses the Harris measure for locating the point of interest, where the determinant of the Hessian matrix is maximum. The SURF descriptor is extracted from the circular region around the point of interest. The SURF is very well suited for tasks in object detection, object recognition, or image retrieval.

3.3.2.4 BRISK features

The binary robust invariant scalable key points (BRISK) [29] algorithm is used to detect rotation- and scale-normalized corner features from an input image. This algorithm is robust to orientation change and finds the perfect key point. The algorithm extracts the salient features from the circular region around the key point. BRISK locates the circular region by rotating the sampling pattern by an angle $\theta = arctan\,2(p_x, p_y)$ around the key point k, where long-distance pairs p_x, p_y are used for computation. It uses fewer sampling points as compared to pairwise comparisons. The BRISK algorithm obtains a high-quality description of an image with low computational cost. Fig. 3.10 shows the visual results of various handcrafted feature extraction techniques. The color histogram and HOG extraction methods obtain the representative pixels from the entire image, whereas the BRISK and SURF techniques obtain features from the local region around the dominated pixels.

3.3.2.5 Deep neural net architectures

Advances in machine learning and deep learning provide an effective way to extract image features without handmade engineering models. Convolutional neural networks (CNNs) are a popular supervised machine learning approach used for learning and extracting features from images. In 2012, Krizhevsky et al. developed a deep CNN (AlexNet) [30] to classify the images of ImageNet and won the Large Scale Visual Recognition Competition (LSVRC). The convolutional neural network learns the best features from an image through multiple convolutions and pooling operations. As the depth of the network grows deeper, the performance of the network increases. The performance of AlexNet is further improved by incorporating the dropout regularization [31] and rectified linear units (ReLUs) [32] activation function. Since the development of AlexNet, many deep CNN architectures have been proposed to reduce the computational load and improve performance, such as OverFeat [33], VGGNet [34], GoogLeNet [35] and ResNet [36]. The details of various deep neural net architectures are given in Table 3.2.

All these CNN models have been trained on the ImageNet database to classify 1000 kinds of objects. These pretrained models have been used in many vision-related systems for extracting the features. The general architecture of the pretrained

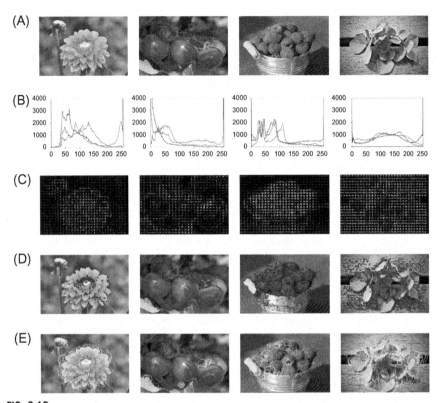

FIG. 3.10

Visual representation of features extracted by engineering methods. (A) Original image (https:/pixabay.com/); (B) RGB color histograms; (C) HOG features; (D) BRISK features and point of interests; (E) SURF features.

Table 3.2 Deep convolutional neural network architectures for feature extraction.

Model	Top-1 accuracy	Top-5 accuracy	Parameters	Depth	Feature shape
VGG16 [34]	0.715	0.901	138.3M	23	$7 \times 7 \times 512$
VGG19 [34]	0.727	0.910	143.6M	26	$7 \times 7 \times 512$
InceptionV3 [35]	0.788	0.944	23.8M	159	$5 \times 5 \times 2048$
ResNet50 [37]	0.759	0.929	25.6M	168	$1 \times 1 \times 2048$
MobileNet [38]	0.665	0.871	4.2M	88	$7 \times 7 \times 1024$
Xception [39]	0.790	0.945	22.9M	126	$1 \times 1 \times 2048$

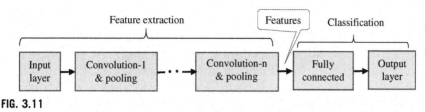

FIG. 3.11

Feature extraction using pretrained CNN models.

DCNN model is shown in Fig. 3.11. Compact feature representations are obtained from the DCNN and this reduced feature representation mitigates computation complexity of clustering or classification systems.

3.3.2.6 Autoencoders

An autoencoder is an unsupervised feature extraction technique that obtains the essential representations from the image. Encoders learn the ideal compressed representation of an image in such a way that the original image can be reconstructed from the compressed form by cotrained decoders. Restricted Boltzmann machine (RBM) autoencoders [40] have been used to extract the multilevel image representations of the image. Since the invention of RBM, many representation learning methods have been proposed to extract features from images based on the deep autoencoder framework. Deep autoencoders are designed to reduce the dimension of image representation by detecting repetitive patterns from the image [41]. The deep autoencoder consists of an encoder part and a decoder part. The encoder down-samples the image into the best level of compression and the decoder tries to reconstruct the image. This process is repeated until the error between the reconstructed image and original image reaches a minimum. After training, the features are extracted from the last block of the encoder part. The architecture model of a deep autoencoder is shown in Fig. 3.12.

The feature maps extracted by the VGG-16 pretrained model and autoencoder model are shown in Fig. 3.13. The visual inspection confirms that the representation learning techniques learn the most contributed pixels during training. Since these models have been trained on a similar domain (natural images), the same pretrained model can be used for fetching the features without fine-tuning.

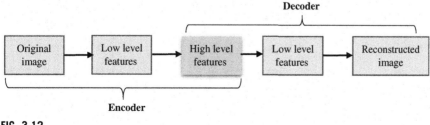

FIG. 3.12

Feature extraction using autoencoder method.

FIG. 3.13

Visual representation of features extracted by representation learning. (A) Original image; (B) Features extracted by VGG16; (C) Features extracted by autoencoder.

3.3.3 Network relation representation

Social network analysis is a study of social relationships in terms of mathematical graph theory. This analysis consists of nodes representing individual images in social media; their ties represent the relationships between the images, such as images commended by the same users, annotated with the same tags, posted by the same users, submitted to the same group, etc. [42–48]. The main objective of social network analysis is the orientation of data. Different terms for nodes and ties are used in various domains, given in Table 3.3.

A graph is also called a network. The example of a graph with three nodes and three edges is illustrated in Fig. 3.14.

The graph element edge may be of two types: directed or undirected. For the directed graph the flow of data will be one-sided, and for the undirected graph, the flow of data will be two-sided. The example of the relation in the graph is given in Fig. 3.15.

Table 3.3 Nodes-ties terms and their domains.

Points	Lines	Domain
Vertices	Edges, arcs	Math
Nodes	Links	Computer Science
Sites	Bonds	Physics
Actors	Ties, relations	Sociology
Atom	Bond	Chemistry

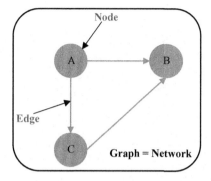

FIG. 3.14

Nodes and edges in graph data.

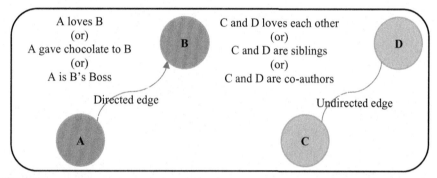

FIG. 3.15

Example of directed and undirected edges.

The edges are capable of holding attributes; the attributes can be weight, rank, or type of edge. The weights in the graph represent the frequency of communication between the two nodes. The ranking may denote best friend, second best friend, and so on. The type of the edge may concern the friendship, relative, coworker, goes for lunch together, etc. The properties may be the betweenness [49, 50], closeness [51, 52], degree [53, 54], etc. An example of a weighted graph is given in Fig. 3.16.

The data representation of the graph can be handled in three different ways: adjacency matrix, edge list, and adjacency list. In these three methods, the adjacency matrix is a straightforward method to vectorize the graph data. In this section, all three data representation methods are discussed.

3.3.3.1 Adjacency matrix

The adjacency matrix [55, 56] is a matrix used to represent finite graphs. The values in the matrix show whether pairs of nodes are adjacent to each other in the graph structure. If the graph is undirected, then the adjacency matrix will be a symmetric one.

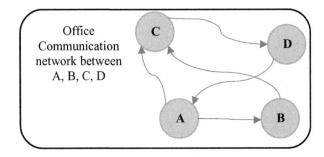

FIG. 3.16

Example graph with edge attributes.

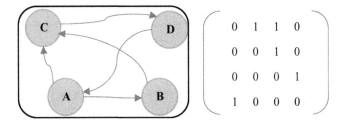

FIG. 3.17

Example of directed graph and the resulting adjacency matrix.

The example graph illustrated in Fig. 3.17 is a directed graph and the resulting adjacency matrix is not a symmetric matrix. In the adjacency matrix, 1 represents that there is an edge from the node A to node B and 0 represents that there is no edge from B to A. Since the graph nodes do not have a self-loop, all the diagonal values are 0.

3.3.3.2 Edge list
The edge list [57] is a method of representing the finest graph in which the detail of the edge is represented as the list. The example of the edge list is illustrated in Fig. 3.18. The length of the list will be equal to the number of edges in the graph.

3.3.3.3 Adjacency list
The adjacency list [46] is a collection of unordered lists used to represent finite graphs. Each list represents the vertex and its neighbors. This method of representation is easier for large and sparse networks. An example representation of the adjacency list is given in Fig. 3.19.

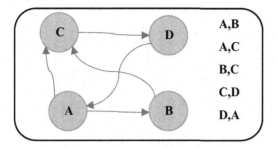

FIG. 3.18

Example of directed graph and the resulting edge list.

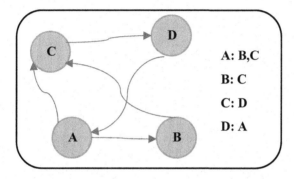

FIG. 3.19

Example of directed graph and the resulting adjacency list.

3.4 Clustering methods

Clustering is one of the important techniques for big data analysis that processes large volumes and different varieties of data, such as structured and unstructured. Clustering methods group the data and present the results with distinct types of meaningful information to analyze it. Clustering is the process of segregating the data points into a number of clusters in which data points in the same clusters are more related to each other than those in other clusters. The clustering can be hard or soft, depending on the membership of the data points to the specific cluster. In hard clustering, the data points are hard-grouped into a cluster. In soft clustering, data points are assigned to clusters based on the probability of their membership in that cluster. The clustering of data can be done based on various models [58] that differ by their organization and the kind of relationship between them, as shown in Fig. 3.20. All the clustering algorithms focus on the 4Vs (volume, variety, velocity, value) of

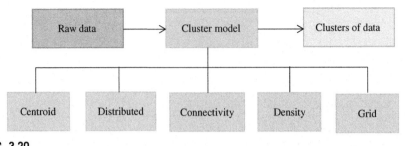

FIG. 3.20

Variants of clustering models.

big data characteristics. The selection of the clustering model is determined by the characteristics of the dataset used and the way the user wants to partition the dataset.

The centroid or partitional model is an iterative clustering model in which each cluster is represented with a cluster centroid that represents the mean of data points in that cluster. Any data point is compared with this cluster centroid to assign a label for it. The number of clusters required for the clustering is to be known beforehand, which means the researcher has prior knowledge of the dataset. The most popular centroid model is a k-means clustering algorithm. ISODATA (Iterative Self Organizing Data Analysis Technique), CLUSTER, WISH, and FORGY are the variations of the k-means algorithm. PAM (Partitioning Around Medoids), CLARA (Clustering LARge Applications), and CLARANS (Clustering Large Applications based on RANdomized Search) are the other centroid-based clustering techniques [59]. CLARANS is a more effective and efficient method as compared to PAM and CLARA. This clustering algorithm is generally suitable for the datasets with spherical-shaped clusters and is not suitable for the complex datasets. Maione et al. have presented a cluster analysis using the PAM clustering algorithm [60] for social data that is capable of handling variables of mixed types and null values. The data related to farmer families from six country cities are used as feature variables to establish profiles for further analysis.

The distributed model creates the clusters with data points that belong to similar statistical distributions (Normal, Gaussian). This model frequently suffers from the problem of overfitting. Two major approaches based on distributed model algorithms, a neural network approach and a statistical approach, are commonly used in data analytics. The expectation-maximization algorithm is one of the examples of this model that uses multivariate normal distributions.

The connectivity model builds the clusters based on the degree of distance between the neighboring data points that show more similarity to each other than the data points lying farther away. This model lacks in scalability while handling big data, but it is very easy to understand. The hierarchical clustering algorithm and its variants fall into this type of clustering model. BIRCH (Balanced Iterative Reducing and Clustering using Hierarchies) [61], CURE (Clustering Using REpresentatives), Chameleon, and ROCK (RObust Clustering using linKs) are the common hierarchical clustering algorithms. This clustering has the limitation that if an incorrect merge or split is done, then the process cannot be revised.

The density-based model creates clusters based on the density of data points grouped by locations. The popular density models are DBSCAN (Density-Based Spatial Clustering of Applications with Noise) and OPTICS (Ordering Points To Identify the Clustering Structure). The grid model works based on the grid structure space related to the data points. Wave-Cluster and STING (STatistical INformation Grid) are typical examples of this clustering model.

Clustering has been used in various domains, such as food science [62, 63], business and customer management, agriculture [64], sports, social media [65–67], and others. One of the most popular applications of clustering is big data analysis in social media. This section reviews some of the clustering techniques employed in social media. A review of the understanding of a large number of multimodal medical datasets using big data analysis methods has been presented [68]. The review stated that the majority of the work in the medical field uses supervised learning but a few semi-or unsupervised methods have been reported for retrieval and understanding. In social media, clustering offers effective organization, search, browsing, and recommendation of data. This supports the clustering of users on Facebook, articles on Google News, videos on YouTube, efficient image retrieval, etc. The clustering techniques can use the features extracted from the input data such as text, image and social relations for similarity computation among them. The photos shared on social media websites may lead to the existence of duplicate images. Wang et al. have evaluated various clustering algorithms [69] that perform mining of duplicate image groups. Web image document stream (WIDS) is a phenomenon that assists in the tracking of topics and the mining of events related to web images shared on the Internet through social media sites. A clustering method has been developed [70] for burst detection that focuses on clustering web images using tags such as text attached to the images.

A comparative analysis has been performed [71] between simple k-means and spectral k-means algorithms for finding the textual similarity to analyze the tweets extracted from Twitter users. A study on different clustering approaches has been performed [59] to analyze their significance in massive datasets to expose their strengths and weakness. Wazarkar [58] has discussed various image-clustering algorithms, substantial feature extraction methodologies, challenges and future research directions for various application domains. The self-organizing map (SOM) neural network [72] is an unsupervised approach that uses competitive learning for both clustering and visualization of data. Some researchers extend the SOM with some modifications to solve high-dimensional unstructured data. A dynamic distributed clustering (DDC) model [73] has been proposed that uses local clustering such as centroid-based k-means and density-based DBSCAN techniques to evaluate large-scale data. A review on big data clustering techniques [74, 75] has been presented to work with bigger datasets through refining their speed and scalability. Since the traditional single-machine clustering algorithms cannot handle an enormous amount of data due to the complexity and computational cost, the importance of multiple machine clustering techniques for handling big data is discussed in their work. An extensive study [76, 77] has been performed on different clustering algorithms that highlights the strengths and limitations of clustering algorithms with respect to the characteristics of big data.

The following section gives an overview of clustering methods and technologies that are commonly used to cluster big data.

3.4.1 The *k*-means clustering algorithm

The *k*-means algorithm solves the clustering problems in an iterative manner that tries to find the local maxima in every iteration. This is one of the simplest unsupervised clustering algorithms that cluster a set of unlabeled data into a specified number of clusters, as shown in Fig. 3.21. The procedure of the *k*-means clustering algorithm is detailed as follows:

Step 1: Specify the number of clusters to cluster the given data points C.
Step 2: Assign each data point $[x_1, x_2, x_3 \ldots x_n]$ to any one of the clusters randomly.
Step 3: Calculate the centroid for each cluster by computing the average of all the data points belonging to that cluster. The centroid (c_i) of the ith cluster in the *k*-means algorithm is represented as the arithmetic mean of the data points present in that cluster:

$$c_i = \frac{1}{T_i} \sum_{x_i \in T_i} x_i \tag{3.4}$$

where $X = [x_1, x_2, x_3 \ldots x_n]$ is the data points in the search space. T_i is the total number of data points present in the *ith* cluster.

Step 4: Reallocate the data points to the nearest cluster centroid based on its minimum Euclidean distance to the cluster centroid. The data point is assigned to the cluster whose distance is minimal to the cluster centroid.

$$\underset{c_i \in C}{argmin} \left[\sum_{j=1}^{N} (x_j - c_{ij})^2 \right] \tag{3.5}$$

where N is the dimension of the features that represent the data point.

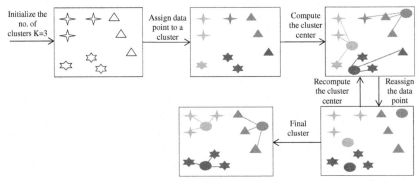

FIG. 3.21

Step-by-step process of *k*-means clustering.

Step 5: After the reallocation of data points, a new centroid (c_i) is recalculated for new clusters.

Step 6: Repeat steps 4 and 5 until the global optima is reached.

The k-means algorithm has a linear complexity $O(n)$ that makes it appropriate for large datasets. The algorithm has its limitations; for example, it is not able to handle higher dimensional nonnumerical attributes or identify nonconvex shaped clusters. This approach is sensitive to data with noise or outliers. Due to some shortcomings present in the k-means, various extensions of classical k-means algorithms have been developed for clustering large datasets. The multiview k-means clustering technique [78] has been proposed to incorporate heterogeneous features for clustering.

3.4.2 Fuzzy *c*-means clustering algorithm

The fuzzy c-means (FCM) algorithm is based on k-means, which partitions the dataset into clusters. Every data point in the dataset has a membership value related to a cluster. The higher the membership value of a data point related to a cluster, the more likely it belongs to that cluster. FCM provides a better result for the overlapped dataset and is relatively superior to the k-means algorithm. The procedure for the FCM clustering algorithm is detailed as follows:

Step 1: Specify the number of clusters C, select the cluster centers randomly, and compute the objective function J.

$$J(U, V) = \sum_{i=1}^{n} \sum_{j=1}^{l} (\mu_{ij})^m \|x_i - c_j\|^2 \tag{3.6}$$

Step 2: Calculate the fuzzy membership μ_{ij} for each data point $[x_1, x_2, x_3 \ldots x_n]$ to the clusters.

$$\mu_{ij} = 1 / \sum_{k=1}^{c} (d_{ij}/d_{ik})^{(2/m-1)} \tag{3.7}$$

where d_{ij} represents the Euclidean distance between ith data and jth cluster center.

Step 3: Compute the cluster center c_j

$$c_j = \frac{\left(\sum_{i=1}^{n} (\mu_{ij})^m x_i \right)}{\left(\sum_{i=1}^{n} \mu_{ij}^{m} \right)}, \forall j = 1, 2, \ldots l \tag{3.8}$$

Step 4: Repeat steps 3 and 4 until the minimum J value is $\|U^{(k+1)} - U^{(k)}\| < \beta$, where k is the iteration step, β is the termination criteria, U is the fuzzy membership matrix.

3.4.3 Mean-shift clustering algorithm

Mean-shift clustering identifies the dense regions of data points using the sliding-window algorithm. This algorithm locates the center points of each cluster by updating the data points for cluster centers to be the mean of the points within the sliding window. The algorithm iteratively shifts the kernel to the higher density region on each step until it reaches convergence. Fig. 3.22 shows the process of mean-shift clustering.

The mean-shift algorithm automatically discovers the number of clusters, in contrast to the k-means. That the cluster centers move towards the points of maximum density is also quite appropriate and fits well in a naturally data-driven sense.

3.4.4 Expectation-maximization algorithm

In Gaussian mixture models (GMMs), the data points are assumed to be Gaussian distributed. The shape of the cluster is described with two parameters, such as mean and standard deviation. An optimization algorithm called expectation-maximization (EM) is applied to identify the parameters of the Gaussian distribution for each cluster. It finds the maximum likelihood parameters of a statistical model. The EM algorithm estimates the unknown model parameters iteratively in two steps, the (1) E step and (2) M step. In the E (expectation) step, the current model parameters are used to calculate the posterior distribution of the latent variables. Based on this value, the data points are fractionally allotted to the clusters. In the M (maximization) step, the fractional assignment is specified by recalculating the model parameters with the maximum likelihood rule. Fig. 3.23 depicts the EM algorithm.

3.4.5 Hierarchical clustering algorithm

Hierarchical clustering algorithms are performed in two approaches, top down or bottom up. In bottom-up algorithms, each data point is treated as a single cluster and then iteratively merges (or agglomerates) with pairs of clusters until all clusters have been merged into a single cluster that holds all data points. The bottom-up hierarchical clustering is also called hierarchical agglomerative clustering (HAC).

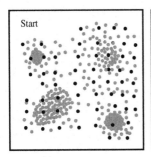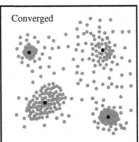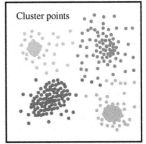

FIG. 3.22

Process of mean-shift clustering.

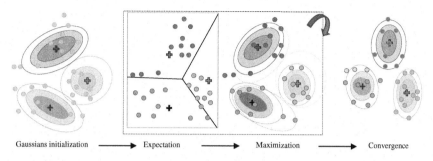

Gaussians initialization ⟶ Expectation ⟶ Maximization ⟶ Convergence

FIG. 3.23

Illustration of expectation-maximization algorithm for Gaussian mixtures.

The top-down approach starts with a set of data points as a single cluster and splits (or divides) the cluster into further clusters until the preferred number of clusters is formed. The top-down approach is also called hierarchical divisive clustering (HDC). The hierarchy of clusters is denoted as a tree called a dendrogram. Fig. 3.24 shows the representation of hierarchical clustering.

The procedure to perform HAC analysis on a dataset is as follows:

Step 1: Find the similarity or dissimilarity between every pair of objects in the dataset by calculating the distance between data points using the similarity measure function.
Step 2: Group the data point into a binary, hierarchical cluster tree. As they are paired into binary clusters, the newly formed clusters are grouped into larger clusters until a hierarchical tree is formed.
Step 3: Determine where to cut the hierarchical tree into clusters and assign all the data points below each cut to a single cluster. This creates a cluster of the data.

Fig. 3.25 shows the process of agglomerative hierarchical clustering with its dendrogram.

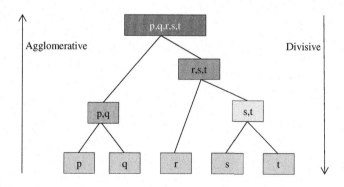

FIG. 3.24

Hierarchical clustering process.

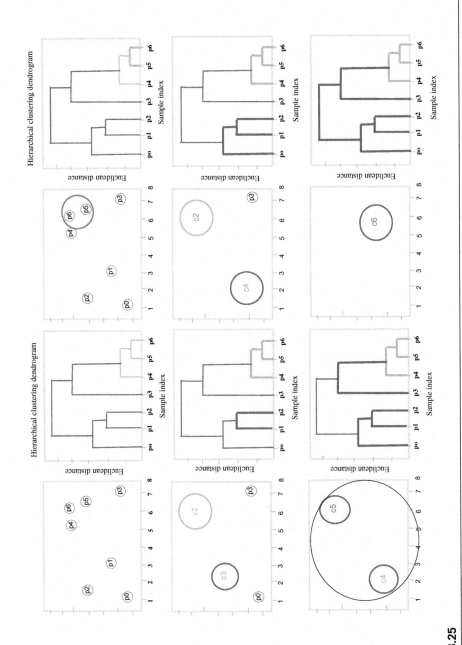

FIG. 3.25

Hierarchical agglomerative clustering.

Unlike other clustering algorithms, this does not require specifying the number of clusters. Also, the algorithm is not sensitive to the choice of distance metric. It is suitable when the underlying data has a hierarchical structure. The hierarchical clustering works slowly with a time complexity of $O(n^3)$.

3.4.6 **DBSCAN algorithm**

DBSCAN is an example of density-based model clustering. The principle of the DBSCAN algorithm works like a liquid flow on a terrain. The liquid starts at a point on the terrain and flows over the terrain where there is the least resistance. The resultant cluster is an area enclosed by the liquid. There are three types of points: core points belong to cluster centers, border points belong to clusters but lie in much less dense parts of them, and noise points do not belong to any cluster. The DBSCAN algorithm is a procedure to discover a given data point's epsilon neighbors. Fig. 3.26 shows the representation of points in DBSCAN.

The procedure for the DBSCAN clustering algorithm is detailed as follows:

Step 1: The algorithm starts with an arbitrary starting data point that has not been visited. The points within the distance "ε" are extracted as neighborhood points.

Step 2: The clustering process starts with "*minPoints*" within this neighborhood and the current data point becomes the first point in the new cluster. Otherwise, it will be labeled as noise. In either case the data point is marked as "visited."

Step 3: All the points within an "ε" distance neighborhood for the first point identified in the new cluster become part of that cluster. This process is repeated for all new points that have been just added to the cluster group with the "ε" distance neighborhood.

Step 4: Repeat steps 2 and 3 until all points are labeled in the cluster as visited.

Step 5: Once the current cluster is formed, select a new unvisited data point and discover further a cluster or noise. Repeat the process until all points are labeled as visited. Finally, each data point has been labeled as either fitting to a cluster or being noise.

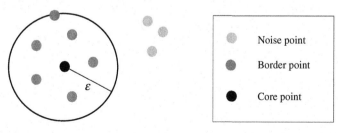

FIG. 3.26

Data point representation in DBSCAN.

3.4.7 **OPTICS algorithm**

The approach of OPTICS is similar to that of DBSCAN and it is an extension of it. DBSCAN is sensitive to parameter setting. OPTICS stores the clustering in two pieces of information such as core distance and reachability distance. The key parameter is *minPoints*, which roughly controls the minimum size of a cluster. The output of OPTICS is not a dataset clustering but a cluster ordering. Data points in a denser cluster are listed closely to each other in the cluster ordering. The data point q is directly density reachable to data point p if q is within the ε distance neighborhood of p and p is a core point.

The procedure for the OPTICS clustering algorithm is detailed as follows:

Step 1: Begin with an arbitrary data point from the input dataset as the current data point p.
Step 2: Retrieve the data points within the ε distance neighborhood of p, determine the core distance, and set the reachability distance to undefined.
Step 3: The current data point p is written to output.
Step 4: If p is not a core point, OPTICS simply moves to the next data point in the Order list or the input space. Data points in the Order list are sorted by the reachability distance from their respective closest core point.

If p is a core point, then for each point q in the ε distance neighborhood of p, OPTICS updates the reachability distance from p and inserts q in the Order list if the q has not been processed yet.
Step 5: The iteration continues until all the data points are fully investigated and the Order list is empty.

3.4.8 **Multiple-machine clustering techniques**

The multiple-machine clustering techniques can run on several machines and have access to more resources. Recently many machine clustering techniques have been developed because they offer faster response and are more flexible in scalability.

a. Parallel clustering: In parallel clustering, large amounts of data are processed in parallel, which achieves results in a reasonable time. The parallel clustering divides the dataset into partitions that will be distributed on different machines. This makes an individual clustering to speed up the calculation and increases scalability. The parallel clustering methods include parallel k-means, parallel fuzzy c-means, etc.
b. MapReduce-based clustering: MapReduce is a task partitioning mechanism that works in distributed execution on a large number of servers. The principle of this clustering is to decompose a task (the map part) into smaller tasks. The tasks are then dispatched to different servers and the results are collected and consolidated (the reduce part).

3.5 Conclusion

Designing a framework for encoding and analyzing social images is very important in the field of social data analytics. This chapter gives an overview of various methods used for mining information from the multimodal social media data along with their metadata, such as tags, comments, images, and social links. The strength and limitations of various feature extraction techniques are explored with respect to social data analytics. The challenges in grouping the vector representation of social images and the capabilities and limitations of various clustering approaches are discussed. Data analysis supports various applications like decision making, predictive analysis, opinion mining, and information retrieval systems. Social media data analytics is a very active research area that still needs more comprehensive studies for evolving a state-of-the-art data analytics model.

References

[1] F. Huang, X. Zhang, Z. Zhao, Z. Li, Y. He, Deep multi-view representation learning for social images, Appl. Soft Comput. 73 (2018) 106–118.

[2] O. Kwon, N. Lee, B. Shin, Data quality management, data usage experience and acquisition intention of big data analytics, Int. J. Inf. Manag. 34 (3) (2014) 387–394.

[3] K. Lyu, H. Kim, Sentiment analysis using word polarity of social media, Wirel. Pers. Commun. 89 (3) (2016) 941–958.

[4] J. Zhao, K. Liu, L. Xu, Sentiment Analysis: Mining Opinions, Sentiments, and Emotions, MIT Press, 2016.

[5] A. Gandomi, M. Haider, Beyond the hype: big data concepts, methods, and analytics, Int. J. Inf. Manag. 35 (2) (2015) 137–144.

[6] N.A. Sakr, A.I. ELdesouky, H. Arafat, An efficient fast-response content-based image retrieval framework for big data, Comput. Electr. Eng. 54 (2016) 522–538.

[7] B. Perozzi, R. Al-Rfou, S. Skiena, DeepWalk: online learning of social representations, in: Proceedings of the 20th ACM SIGKDD International Conference on Knowledge Discovery and Data Mining—KDD '14, 2014, pp. 701–710, https://doi.org/10.1145/2623330.2623732.

[8] S. Stieglitz, M. Mirbabaie, B. Ross, C. Neuberger, Social media analytics – challenges in topic discovery, data collection, and data preparation, Int. J. Inf. Manag. 39 (2018) 156–168.

[9] Y. Gong, Q. Ke, M. Isard, S. Lazebnik, A multi-view embedding space for modeling internet images, tags, and their semantics, Int. J. Comput. Vis. 106 (2) (2014) 210–233.

[10] M. Mhatre, D. Phondekar, P. Kadam, A. Chawathe, K. Ghag, Dimensionality reduction for sentiment analysis using pre-processing techniques, in: 2017 International Conference on Computing Methodologies and Communication (ICCMC), IEEE, 2017, pp. 16–21.

[11] A.L. Mullen, K. Benoit, O. Keyes, D. Selivanov, J. Arnold, Fast, consistent tokenization of natural language text, J. Open Source Softw. 3 (23) (2018) 655, https://doi.org/10.21105/joss.00655.

[12] J. Jadav, C. Tappert, M. Kollmer, A.M. Burke, P. Dhiman, Using text analysis on web filter data to explore K-12 student learning behavior, in: 2016 IEEE 7th Annual Ubiquitous Computing, Electronics and Mobile Communication Conference, UEMCON 2016, 2016, , pp. 1–5, https://doi.org/10.1109/UEMCON.2016.7777882.

[13] F.R. Lucini, F.S. Fogliatto, G.J. Giovani, J.L. Neyeloff, M.J. Anzanello, R.S. de Kuchenbecker, B.D. Schaan, Text mining approach to predict hospital admissions using early medical records from the emergency department, Int. J. Med. Inf. 100 (2017) 1–8, https://doi.org/10.1016/j.ijmedinf.2017.01.001.

[14] J. Yang, E. Kim, M. Hur, S. Cho, M. Han, I. Seo, Knowledge extraction and visualization of digital design process, Expert Syst. Appl. 92 (2018) 206–215, https://doi.org/10.1016/j.eswa.2017.09.002.

[15] J. Camacho-Collados, M.T. Pilehvar, On the Role of Text Preprocessing in Neural Network Architectures: An Evaluation Study on Text Categorization and Sentiment Analysis, https://doi.org/arXiv:1707.01780v3, 2017.

[16] R. Nokhbeh Zaeem, M. Manoharan, Y. Yang, K.S. Barber, Modeling and analysis of identity threat behaviors through text mining of identity theft stories, Comput. Secur. 65 (2017) 50–63, https://doi.org/10.1016/j.cose.2016.11.002.

[17] J. Singh, V. Gupta, A systematic review of text stemming techniques, Artif. Intell. Rev. 48 (2) (2017) 157–217, https://doi.org/10.1007/s10462-016-9498-2.

[18] B. Altınel, M.C. Ganiz, Semantic text classification: a survey of past and recent advances. Inf. Process. Manag. 54 (6) (2018) 1129–1153, https://doi.org/10.1016/j.ipm.2018.08.001.

[19] S. Bansal, C. Gupta, A. Arora, User tweets based genre prediction and movie recommendation using LSI and SVD, in: 2016 9th International Conference on Contemporary Computing, IC3 2016, 2017, https://doi.org/10.1109/IC3.2016.7880220.

[20] P. Barnaghi, P. Ghaffari, J.G. Breslin, Opinion mining and sentiment polarity on twitter and correlation between events and sentiment, in: Proceedings—2016 IEEE 2nd International Conference on Big Data Computing Service and Applications, BigDataService 2016, 2016, pp. 52–57, https://doi.org/10.1109/BigDataService.2016.36.

[21] L. Xu, C. Jiang, Y. Ren, H.H. Chen, Microblog dimensionality reduction – a deep learning approach, IEEE Trans. Knowl. Data Eng. 28 (7) (2016) 1779–1789, https://doi.org/10.1109/TKDE.2016.2540639.

[22] W. Yang, Z. Fang, L. Hui, Study of an improved text filter algorithm based on trie tree, in: Proceedings – 2016 IEEE International Symposium on Computer, Consumer and Control, IS3C 2016, 2016, pp. 594–597, https://doi.org/10.1109/IS3C.2016.153.

[23] D.G. Lowe, Distinctive image features from scale-invariant keypoints, Int. J. Comput. Vis. 60 (2) (2004) 91–110.

[24] H. Bay, A. Ess, T. Tuytelaars, L. Van Gool, Speeded-up robust features (SURF), Comput. Vis. Image Underst. 110 (3) (2008) 346–359.

[25] Z. Mehmood, F. Abbas, T. Mahmood, M.A. Javid, A. Rehman, T. Nawaz, Content-based image retrieval based on visual words fusion versus features fusion of local and global features, Arab. J. Sci. Eng. (2018) 1–20.

[26] T.-W. Chen, Y.-L. Chen, S.-Y. Chien, Fast image segmentation based on K-means clustering with histograms in HSV color space, in: 2008 IEEE 10th Workshop Multimedia Signal Processing, 2008, pp. 322–325.

[27] G.-H. Liu, J.-Y. Yang, Content-based image retrieval using color difference histogram, Pattern Recogn. 46 (1) (2013) 188–198.

[28] J. Arróspide, L. Salgado, M. Camplani, Image-based on-road vehicle detection using cost-effective histograms of oriented gradients, J. Vis. Commun. Image Represent. 24 (7) (2013) 1182–1190.

[29] S. Leutenegger, M. Chli, R.Y. Siegwart, BRISK: binary robust invariant scalable keypoints, in: 2011 IEEE International Conference on Computer Vision (ICCV), 2011, pp. 2548–2555.

[30] A. Krizhevsky, I. Sutskever, G.E. Hinton, Imagenet classification with deep convolutional neural networks, Adv. Neural Inf. Proces. Syst. (2012) 1097–1105.

[31] G.E. Hinton, N. Srivastava, A. Krizhevsky, I. Sutskever, R.R. Salakhutdinov, Improving neural networks by preventing co-adaptation of feature detectors, 2012. ArXiv Preprint ArXiv:1207.0580.

[32] V. Nair, G.E. Hinton, Rectified linear units improve restricted boltzmann machines, in: Proceedings of the 27th international conference on machine learning (ICML-10), 2010, pp. 807–814.

[33] P. Sermanet, D. Eigen, X. Zhang, M. Mathieu, R. Fergus, Y. LeCun, Overfeat: Integrated recognition, localization and detection using convolutional networks, 2013. ArXiv Preprint ArXiv:1312.6229.

[34] K. Simonyan, A. Zisserman, Very deep convolutional networks for large-scale image recognition, 2014. ArXiv Preprint ArXiv:1409.1556.

[35] C. Szegedy, W. Liu, Y. Jia, P. Sermanet, S. Reed, D. Anguelov, … A. Rabinovich, Going deeper with convolutions, CoRR abs/1409.4842, http://Arxiv.Org/Abs/1409.4842, 2014.

[36] K. He, X. Zhang, S. Ren, J. Sun, Deep residual learning for image recognition, in: Proceedings of the IEEE Conference on Computer Vision and Pattern Recognition, 2016, pp. 770–778.

[37] Q. He, J. Johnston, J. Zeitlinger, K. City, ChIP-nexus enables improved detection of in vivo transcription factor binding footprints, Nat. Biotechnol. 33 (4) (2015) 395–401, https://doi.org/10.1038/nbt.3121.

[38] A.G. Howard, M. Zhu, B. Chen, D. Kalenichenko, W. Wang, T. Weyand, … H. Adam, Mobilenets: Efficient convolutional neural networks for mobile vision applications, 2017. ArXiv Preprint ArXiv:1704.04861.

[39] F. Chollet, Xception: Deep Learning With Depthwise Separable Convolutions, ArXiv Preprint, 2017, pp. 1610–2357.

[40] G.E. Hinton, R.R. Salakhutdinov, Reducing the dimensionality of data with neural networks, Science 313 (5786) (2006) 504–507.

[41] Y. Wang, H. Yao, S. Zhao, Auto-encoder based dimensionality reduction, Neurocomputing 184 (2016) 232–242.

[42] F. Battiston, V. Nicosia, V. Latora, Structural measures for multiplex networks, Phys. Rev. E Stat. Nonlinear Soft Matter Phys. 89 (3) (2014) 1–16, https://doi.org/10.1103/PhysRevE.89.032804.

[43] S. Boccaletti, G. Bianconi, R. Criado, C.I. del Genio, J. Gómez-Gardeñes, M. Romance, … M. Zanin, The structure and dynamics of multilayer networks, Phys. Rep. 544 (1) (2014) 1–122, https://doi.org/10.1016/j.physrep.2014.07.001.

[44] M.A. Rodriguez, J. Shinavier, Exposing multi-relational networks to single-relational network analysis algorithms, J. Inf. 4 (1) (2010) 29–41, https://doi.org/10.1016/j.joi.2009.06.004.

[45] L. Solá, M. Romance, R. Criado, J. Flores, A. García del Amo, S. Boccaletti, Eigenvector centrality of nodes in multiplex networks, Chaos 23 (3) (2013) 1–11, https://doi.org/10.1063/1.4818544.

[46] X. Wang, J. Liu, A layer reduction based community detection algorithm on multiplex networks, Physica A 471 (2017) 244–252, https://doi.org/10.1016/j.physa.2016.11.036.

[47] K. Xu, K. Zou, Y. Huang, X. Yu, X. Zhang, Mining community and inferring friendship in mobile social networks, Neurocomputing 174 (2016) 605–616, https://doi.org/10.1016/j.neucom.2015.09.070.

[48] W.W. Zachary, An information flow model for conflict and fission in small groups, J. Anthropol. Res. 33 (4) (1977) 452–473, https://doi.org/10.1086/jar.33.4.3629752.

[49] R. Kanawati, Seed-centric approaches for community seed-centric algorithms : a classification study, in: International Conference on Social Computing and Social Media SCSM 2014: Social Computing and Social Media, 2014, pp. 197–208.

[50] Newman, M. Girvan, Finding and evaluating community structure in networks, Phys. Rev. E 69 (2) (2004) 26113, https://doi.org/10.1103/PhysRevE.69.026113.

[51] B. Andreopoulos, A. An, X. Wang, M. Schroeder, A roadmap of clustering algorithms: finding a match for a biomedical application, Brief. Bioinform. 10 (3) (2009) 297–314, https://doi.org/10.1093/bib/bbn058.

[52] R. Xu, Survey of clustering algorithms for MANET, IEEE Trans. Neural Netw. 16 (3) (2005) 645–678, https://doi.org/10.1109/TNN.2005.845141.

[53] T. Nepusz, A. Petroczi, L. Negyessy, F. Bazsso, Fuzzy communities and the concept of bridgeness in complex networks, Phys. Rev. E Stat. Nonlinear Soft Matter Phys. 77 (1) (2008) 1–13, https://doi.org/10.1103/PhysRevE.77.016107.

[54] G. Song, Y. Li, X. Chen, X. He, J. Tang, Influential node tracking on dynamic social network: an interchange greedy approach, IEEE Trans. Knowl. Data Eng. 29 (2) (2017) 359–372, https://doi.org/10.1109/TKDE.2016.2620141.

[55] N. Barbieri, F. Bonchi, G. Manco, Efficient methods for influence-based network-oblivious community detection, ACM Trans. Intell. Syst. Technol. 8 (2) (2016), https://doi.org/10.1145/2979682.

[56] J. Xie, M. Chen, B. Szymanski, Labelrank T: incremental community detection in dynamic networks via label propagation, in: Proc. DyNetMM Workshop at the SIGMOD/PODS Conference, New York, NY, 22–27 June, 2013, pp. 25–32, https://doi.org/10.1145/2489247.2489249.

[57] Y. Xu, H. Xu, D. Zhang, Y. Zhang, Finding overlapping community from social networks based on community forest model, Knowl.-Based Syst. 109 (2016) 238–255, https://doi.org/10.1016/j.knosys.2016.07.007.

[58] S. Wazarkar, B.N. Keshavamurthy, A survey on image data analysis through clustering techniques for real world applications, J. Vis. Commun. Image Represent. 55 (2018) 596–626.

[59] P. Nerurkar, A. Shirke, M. Chandane, S. Bhirud, Empirical analysis of data clustering algorithms, Procedia Comput. Sci. 125 (2018) 770–779, https://doi.org/10.1016/j.procs.2017.12.099.

[60] C. Maione, D.R. Nelson, R.M. Barbosa, Research on social data by means of cluster analysis, Appl. Comput. Inf. 15 (2) (2018) 153–162.

[61] T. Zhang, R. Ramakrishnan, M. Livny, BIRCH: an efficient data clustering method for very large databases, in: SIGMOD '96 Proceedings of the 1996 ACM SIGMOD International Conference on Management of Data, vol. 25, 1996, pp. 103–114.

[62] C. Maione, B.L. Batista, A.D. Campiglia, F. Barbosa Jr., R.M. Barbosa, Classification of geographic origin of rice by data mining and inductively coupled plasma mass spectrometry, Comput. Electron. Agric. 121 (2016) 101–107.

[63] C. Maione, E.S. de Paula, M. Gallimberti, B.L. Batista, A.D. Campiglia, F. Barbosa Jr., R.M. Barbosa, Comparative study of data mining techniques for the authentication of organic grape juice based on ICP-MS analysis, Expert Syst. Appl. 49 (2016) 60–73.

[64] A. Mucherino, P. Papajorgji, P.M. Pardalos, A survey of data mining techniques applied to agriculture, Oper. Res. 9 (2) (2009) 121–140.

[65] G. Barbier, H. Liu, Data mining in social media, in: Social Network Data Analytics, Springer, 2011, pp. 327–352.

[66] P. Gundecha, H. Liu, Mining social media: a brief introduction, in: New Directions in Informatics, Optimization, Logistics, and Production, Informs, 2012, pp. 1–17.

[67] C. Papagiannopoulou, V. Mezaris, Concept-based image clustering and summarization of event-related image collections, in: Proceedings of the 1st ACM International Workshop on Human Centered Event Understanding from Multimedia, 2014, pp. 23–28.

[68] H. Müller, D. Unay, Retrieval from and understanding of large-scale multi-modal medical datasets: a review, IEEE Trans. Multimedia 19 (9) (2017) 2093–2104.

[69] B. Wang, Z. Li, M. Li, W.-Y. Ma, Large-scale duplicate detection for web image search, in: 2006 IEEE International Conference on Multimedia and Expo, 2006, pp. 353–356.

[70] S. Tamura, K. Tamura, H. Kitakami, K. Hirahara, Clustering-based burst-detection algorithm for web-image document stream on social media, in: 2012 IEEE International Conference on Systems, Man, and Cybernetics (SMC), 2012, pp. 703–708.

[71] K. Singh, H.K. Shakya, B. Biswas, Clustering of people in social network based on textual similarity, Perspect. Sci. 8 (2016) 570–573.

[72] O. Kurasova, V. Marcinkevicius, V. Medvedev, A. Rapecka, P. Stefanovic, Strategies for big data clustering, in: 2014 IEEE 26th International Conference on Tools with Artificial Intelligence (ICTAI), 2014, pp. 740–747.

[73] M. Bendechache, M.-T. Kechadi, N.-A. Le-Khac, Efficient large scale clustering based on data partitioning, in: 2016 IEEE International Conference on Data Science and Advanced Analytics (DSAA), 2016, pp. 612–621.

[74] A.S. Shirkhorshidi, S. Aghabozorgi, T.Y. Wah, T. Herawan, Big data clustering: a review, in: International Conference on Computational Science and Its Applications, 2014, pp. 707–720.

[75] B. Zerhari, A.A. Lahcen, S. Mouline, Big data clustering: Algorithms and challenges, in: Proceedings of International Conference on Big Data, Cloud and Applications (BDCA'15), 2015.

[76] A. Fahad, N. Alshatri, Z. Tari, A. Alamri, I. Khalil, A.Y. Zomaya, … A. Bouras, A survey of clustering algorithms for big data: taxonomy and empirical analysis, IEEE Trans. Emerg. Top. Comput. 2 (3) (2014) 267–279.

[77] T. Sajana, C.M.S. Rani, K.V. Narayana, A survey on clustering techniques for big data mining, Indian J. Sci. Technol. 9 (3) (2016), https://doi.org/10.17485/ijst/2016/v9i3/75971.

[78] X. Cai, F. Nie, H. Huang, Multi-view K-means clustering on big data, in: IJCAI'13 Proceedings of the Twenty-Third International Joint Conference on Artificial Intelligence, 2013, pp. 2598–2604.

A novel framework for quality care in assisting chronically impaired patients with ubiquitous computing and ambient intelligence technologies

4

Deva Priya Isravel[a], Diana Arulkumar[a], Kumudha Raimond[a], Biju Issac[b]

[a]*Department of Computer Science and Engineering, Karunya Institute of Technology and Sciences, Coimbatore, India*
[b]*Department of Computer and Information Sciences, Northumbria University, Newcastle, United Kingdom*

4.1 Introduction

There is an increasing trend in the occurrence of chronic disease throughout the world. Studies have shown that there is a tremendous need for technology in providing management and preventive support for patients. The traditional clinical processes, monitoring mechanisms, and surveillance systems are inadequate for providing quality care [1]. A significant proportion of the elderly population suffers from age-related health issues. Providing quality healthcare to chronic patients has become a difficult task due to the lack of timely adequate information. Because of technological advances, these problems can be overcome through ubiquitous, intelligent, and pervasive connected systems [2].

An especially wide range of challenging tasks is involved in caring for chronically ill patients. Internet-based or mobile support has proven to be effective. However, patients with chronic illness with cognitive impairment or reduced self-management find it hard to use the technology. Even elderly patients who are not familiar with the Internet or mobile technology find it hard to use [3].

With the advancement of ubiquitous technology and ambient intelligence (AmI), providing quality care is within reach. The innovations of cloud computing and artificial intelligence have paved the way for adoption of ubiquitous computing (UC). The rise in the use of context-aware appliances has improved the progress of the healthcare industry. With UC, multiple microprocessors and sensors can be embedded into objects and they can be wirelessly connected [4, 5]. AmI is a promising technology where embedded intelligent tools facilitate communication with persons and are aware of context, such as location, changes in the environment, and presence of humans. These tools are very sensitive, responsive, and adaptive to satisfy a patient's needs. They can record occurrences in the environment like gestures, emotions, habits, etc. and communicate the data for analysis, as well as take corrective actions [6]. The advent of smart computing devices has led to AmI, which enables these devices to exhibit cognitive capabilities by being mobile, observant, reasoning, and proactively taking measures to adapt to changing situations [7]. AmI facilitates assisted living environments [8]. They record the environment and communicate data that can be used to take corrective actions. It helps to augment a typical hospital facility with smart features by providing the patients and staff with crucial and relevant assistance without the need for human involvement.

Improving the quality of healthcare and providing cost-effective solutions is a requirement of smart healthcare applications. The aim of applying UC is to help patients manage their disease and enable them to adhere to medical treatments. The beneficiaries of this technology are patients, doctors, and hospital staff. There are a lot of challenging tasks involved in taking care of chronically ill patients. Most often chronic illness leads to physical disability. These are patients who have a persistent disease and often need supportive services. Many have self-management disabilities requiring constant monitoring and assistance to do even the simplest task. The patients must be continuously monitored for vital signs to diagnose any arising problem and provide treatment. In emergency situations, timely help must be offered to patients with telemedicine and health-related guidelines. In essence, smart healthcare facilitates a patient-driven operations model rather than the existing doctor-centric system for better handling of chronic patients [2].

The development of the Internet of Things (IoT) has contributed to quality and convenient medical services. A collection of a few smart sensors implanted into or placed on the patient's body or in a wearable device could observe vital signs and help in handling an emergency or any critical and chronic conditions. Through this technology, the participants would include patients, nurses, doctors, smart appliances, clinical equipment, and medicines [9].

In this chapter, a novel framework is proposed with UC and ambient intelligence to provide professional and quality care to chronic patients with technologies such as edge, fog, cloud computing, big data, IoT, and cognitive analytics. In Section 4.2, the theoretical perspective of the various technologies is presented. Section 4.3 covers the various research challenges in the healthcare system. Related works which have already been carried out to address patient requirements are detailed in Section 4.4. In Section 4.5, a novel framework for the self-managed environment for an assisted patient living using UC and ambient intelligence is explained. Section 4.6 presents the conclusion.

4.2 **The theoretical perspective of ubiquitous computing**

This section gives an overview of the evolution of the healthcare system over the years. After the advancements in IT and networking technology, the quality of health services offered has improved tremendously. Also, the various technologies that contribute to UC are elaborated. Fig. 4.1 depicts the evolution of the healthcare system.

The traditional healthcare system required the patient to be available wherever the physician was present, and the quality of treatment offered was restricted and limited. Health needs were numerous but resource availability and accessibility were limited. The need grew for improvements in the quality of the healthcare system, as it is very much essential for any community. Hospital recordkeeping was improved with desktop computers for storing the health-related documents of patients, making access to the patient documents easier. At first, these documents were accessible only by the hospital unit.

The productivity of the health industry was improved with less labor and reduced time, as the large-scale health system benefited from the development of Windows-based IT solutions. The introduction of the Internet to all users brought about a huge change in the health industry. The hospital infrastructure was upgraded by establishing wired connectivity between clinical equipment and computers. This improved collaboration among the patients, doctors, and users of the system. The communication processes improved significantly, but there were challenges in maintaining the privacy and dissemination of data. This paved the way for telemedicine, wherein diagnosis and treatment can be carried out from a distance, allowing services to be offered beyond hospital boundaries.

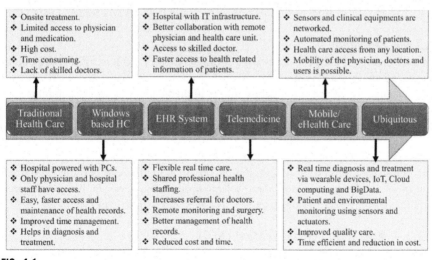

FIG. 4.1

Evolution of healthcare system.

The next leap to improve healthcare was the boom in wireless connectivity. Every hospital network environment now has wireless connectivity. With mobile devices and PDAs, access to online e-health records became more convenient and readily available with proper authentication. The current system of health services offered is called ubiquitous healthcare, wherein many technologies work together to provide better healthcare services. Wearable devices in combination with a variety of sensors make allow the continuous observation of patients in terms of symptoms, vital signs, and behavioral patterns. The observed data is sent as graphics images or as a measurement for initiating corrective treatment. Image archives require cloud technology for storage and cognitive analytics for interpreting the collected data, which can then be used for diagnosis and further treatment.

4.2.1 Wireless technologies

There are various wireless technologies that facilitate the realization of ubiquitous technology for smart and quality healthcare. The following section explains in detail the most commonly used technology.

4.2.1.1 Radio frequency identification (RFID)

The automatic identification of objects and individuals is an essential component of any application. RFID is a technology that identifies and localizes objects from a distance and it can be used in monitoring patients [4]. The RFID chip can store a lot of information, which can be transferred to the cloud for further processing. It is a relatively simple and cost-effective technology that can be used for tracking chronically impaired patients, monitoring medication usage, etc. from a remote location [10]. Due to its high positional accuracy, it can be used as a tracking tool [11].

4.2.1.2 Bluetooth

This technology is used for short-range communication [12]. It is a low energy consuming technology and is ubiquitously deployed for monitoring purposes. Bluetooth was designed to replace serial cables and is implemented in medical equipment to wirelessly transmit medical images as signals to the terminal system, where they are processed to obtain useful information for treatment [11].

4.2.1.3 ZigBee

This is a specification for low-cost, low-power digital radios and has found application in areas like home automation, telecommunication services, healthcare, and remote control. It can transfer data at a low rate and within 10m [12].

4.2.1.4 Near field communication (NFC)

NFC is a technology that transfers data between devices in close proximity using RF signals. It is used in various applications like keyless access, credit cards, E-wallets, smart ticketing, wearable devices, smart tags for medical applications, etc. It is one of the most convenient technologies that operates at low power and low frequency range. It can be used for powering implantable electronic devices. However, there is a constant demand for reductions in size and complexity for medical devices because they come with a special set of requirements and design constraints [13]. Wireless NFC tags can be used with a wide range of applications like temperature sensing, biochemical sensing [14], monitoring wound infection condition [15], patient tracking, prescribing medicines, etc.

4.2.1.5 Sensors

A sensor is a device or module used to sense, detect, or measure a physical phenomenon and respond to the physical environment. A huge range of sensor types are currently in use for various applications. These include magnetic, radio, temperature, humidity, flow, blood pressure, pulse oximetry, ultrasonic, air bubble detection, force, pH, proximity, optical, position, chemical, magnetic switch, PIR, speed, and motion sensors, among many other. All these intelligent sensors collect and send a huge amount of data to the computing system via any of the wireless communication technologies so the data can be analyzed and action taken on the events observed. A collection of different sensors can be deployed for remote monitoring of chronic patients, assisting them with their needs, observing vital signs for providing clinical treatments, and monitoring the hospital surroundings.

4.2.2 Wearable technologies

Wearable devices (WDs) can be useful in early detection and management of medical conditions. Advanced wearable devices aid physicians and surgeons in knowing the physiological parameters of patients and assisting their patients in real time, based on transmitted data [16]. Studies have shown that elderly populations are more prone to benefits from using WDs. These WDs can measure the vital signs of patients and health conditions such as body and skin temperature, arterial blood pressure, heart rate, respiration rate, electrocardiogram (ECG), etc., as well as monitor physical movement. All these WDs are available with various technological capabilities, benefits, and costs. The patients should have the skills to use this technology. Patients with a chronic condition should be offered restorative treatment when they fall ill. Any delays in receiving full treatment could worsen the patient's health condition, so the WDs can reduce the health deterioration rate as they collect accurate and real-time data [17].

4.2.3 Human activity recognition

Human activity recognition (HAR) plays a significant role in monitoring human activities and has attracted significant research interest due to its applications in health monitoring and patient rehabilitation. The activity recognition devices include cameras, touch sensing devices, text entry devices such as special keyboards, chord keyboards, digesting tablet, voice recognition devices, and devices for understanding gestures and facial expressions. As the system is designed for facilitating chronically ill patients, who may not be in a position to operate the devices, facial expressions and gestures have to be studied to understand the patient needs in real time. Many companies have introduced human-computer interaction devices based on the latest technologies that can be extended for the use of chronically ill patients. Some of the devices are as follows:

- Leap Motion, Inc. is an American company that manufactures a sensor device that supports hand and finger motions without requiring any hand contact or touching.
- Google Glass is a wearable computer featuring a head-mounted display in the form of eyeglasses. It allows users to access smartphone apps by voice commands.
- Eyegaze Edge empowers the user to communicate and interact with the world by looking at the control keys or cells displayed on a screen and generate speech by typing a message.

4.2.4 Internet of Things

The IoT has brought tremendous improvements into the healthcare system by means of remote monitoring, patient tracking, inventory management, ensuring the availability and accessibility of clinical equipment, drug management, among many others. It has helped to modernize the tech-driven hospital and medical facility. IoT applications can range from simple fitness aids to chronic disease management technology [2]. There are certain healthcare applications that do not require diagnosis or treatment, but there are other applications that can target chronically impaired patients and provide advanced medical care through patient monitoring, vital signs and disease detection, interaction with medical experts, and even remote treatment. This technology allows facilities and resources to be networked in order to acquire real-time data and facilitate the decision-making process. With smart healthcare, the data collected can be analyzed through data analytics, and an emergency situation can be handled and responded to in a more timely manner if the resources can be managed and controlled appropriately [18].

The generic IoT architecture has five layers of architecture: business, application, middleware layer, network, and perception layer [19]. The middleware layer supports ubiquitous computing facilities for storage and retrieval of the large volume of data. Fig. 4.2 shows the layers in the IoT architecture.

- **Perception layer:** This layer corresponds to data acquisition from the patient and the environment using sensors and medical instruments. It captures movements, gestures, and positions of the patient. The captured data is transmitted to the middleware layer's database via the network layer.
- **Network layer:** This represents the connectivity offered by wired or wireless networks along with the various technologies such as WiFi, Bluetooth, Zigbee, etc.
- **Middleware layer:** The data log is collected in real time and stored in the database. It processes the information based on the available data for making automatic decisions. This layer also assists in short, medium, and long-term data analysis.

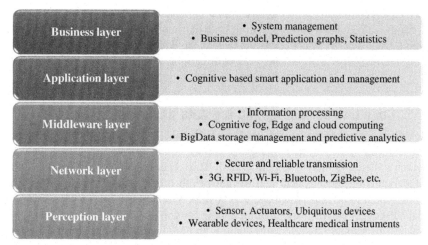

FIG. 4.2

IoT architecture.

- **Application layer:** This layer represents the various tools and frameworks that can be used for cognitive-based smart analysis, decision making, and management.
- **Business layer:** This layer helps to visualize and gain more insight into the real-time solution provided by the application layer by means of charts, graphs, and statistics.

4.2.5 **Cloud computing**

As IoT and other technologies generate a high volume of patient-related data (24×7) from sensors and other medical-related devices, the process must support good storage infrastructure and retrieval for short- and long-term data analysis. This requirement has led to a significant milestone in UC-based medical services, as such resources are mostly being provided by cloud service providers. Here the computation and storage capacity exist to enable devices to make "smart" decisions without human intervention. Both these technologies are inseparable and are often used together to improve performance in developing healthcare solutions with more intelligent analysis and prediction capabilities. But the cloud model is not suitable for time-critical medical cases when internet connectivity is poor. This is especially true in telemedicine applications and in chronic patient care, where small delays of milliseconds due to long round-trip times to the cloud server could have fatal consequences.

Edge and fog computing represent an essential paradigm shift towards a hierarchical cloud architecture, as they extend cloud computing and provide storage, computing infrastructure, and a data analytics platform closer to the endpoints. These new cloud infrastructures play a significant role in IoT applications that requires low latency and low cost.

Edge/fog computing provides many benefits, such as increased network capacity, less bandwidth, and real-time operation, as all data analysis is carried out in real-time fashion. Data security is not compromised, as the data remains in the local network during the real-time analysis and alert notifications.

4.2.6 Cognitive computing and big data analytics

The recent advancements in the areas of artificial intelligence, machine learning, deep learning, and natural language processing have resulted in a new field of research called cognitive computing [20]. This emerging field will assist doctors in understanding the physical and mental health of the patient more completely, enabling them to offer personalized diagnosis and treatments. Such a new approach to data discovery and decision making is called cognitive analytics. It overcomes the limitation of human cognition in handling and analyzing big data from a diverse range of healthcare sources, such as data from distributed ubiquitous devices, sensors, medical equipment, etc., to predict the status of the patient in real time.

4.2.7 Ambient intelligence

Ambient intelligence is another new multidisciplinary technology that provides assisted living solutions for elderly people. Ambient assisted living (AAL) encompasses many healthcare-related applications such as medications reminders, fall detection systems, and monitoring activities of daily living, to securing people from life-threatening situations and improving their wellness. It also helps to monitor and communicate the health status of the elderly patients to family members so that rapid assistance can be provided when needed. AAL technology assists elderly persons to live a more independent life, thereby also providing a needed service to the caregivers, friends, and family [21]. This technology incorporates environmental awareness and enables one to have access to the direct natural and intuitive interaction of the user with applications and services [6]. A great variety of sensors are commonly used in AAL, such as temperature sensors, photosensors, pressure pads, water flow sensors, infrared motion sensors, power/current sensors, force sensors, smoke/heat sensors, biosensors, accelerometers, gyroscope light sensors, proximity sensors, audio sensors, cameras, barometers, heart rate sensors, GPS, and magnetic sensors.

4.3 Research challenges

From an extensive study of the literature, we can understand that there are several challenges that require immediate attention and focus in order to realize the full potential of ubiquitous healthcare. This section presents the major challenges and hurdles faced by the e-healthcare system.

4.3.1 **Context awareness**

Context awareness is the ability of the system to comprehend the patient's vital signs and symptoms and then interpret the sensed data to intelligently respond to the situation. The observed and collected data must be correlated with the current activity to make smart decisions in offering health services. Because there is a great deal of information gathered with respect to behavioral, psychological, environmental and clinical data, processing this data is challenging [22]. There are still open issues that need to be researched for arriving at an optimal and efficient ubiquitous healthcare system.

4.3.2 **Security**

Another major challenge faced in handling the patient's health information is the security and privacy of the information. Because of the sensitive nature and need for confidentiality of the personal records of patients, they must be protected against attacks and vulnerabilities. To ensure the integrity of the data flowing across the network, proper authentication must be done. The patient's records contain personal information like personal identities, family status, financial information, drug and medical dependency data, medical diagnosis and treatment details, and history of health-related matters. Therefore, maintaining privacy while making these data available to different forums must be addressed. Uncontrolled access may lead to misuse of data and substantial threats. With the ubiquitous system, the information is subject to copying, deleting, modification, and misuse. If these important and private data are given to unauthorized third parties, the patient's privacy is violated. Therefore, still, much research is required to ensure trust, privacy, and security.

4.3.3 **Heterogeneity**

The next challenge is to deal with heterogeneity. In order to provide care for the chronically ill patient, a wide variety of devices like sensors, wearable devices, web cameras, RFID tags, clinical equipment, smart devices, etc. are used. They all have to cooperatively work together to provide on-site and remote assistance services. Because of the distributed and heterogeneous nature of the system, the exchange of data in real time with accuracy and consistency is difficult and complex [23]. Though several studies have been carried out to simplify the solutions, further development is still required to address scalability and reliability.

4.3.4 **Data management**

Data management is a cumbersome and expensive task because it involves collecting a vast amount of data from hospitals, patients, doctors, communities, rehabilitation centers, and homes via wearable devices, computing devices, clinical equipment, IoT devices, mobile apps, sensors, and so on [18]. Theoretically, all collected data are stored on servers. Even though the cost of storage has reduced considerably, collection, storage and management of this huge volume of data is challenging as information is updated from various automated devices in real time and critical decisions have to be made simultaneously.

4.3.5 Scalability

Scalability is another challenge that is faced by the health industry. As the size of a network grows, it takes a huge investment from the healthcare organization to expand both wired and wireless connectivity throughout the hospital facility. With mobility, the challenge further intensifies because now communication is not within four walls but across both far and near locations. With smart healthcare, a lot of information is gathered from many sensors, actuators, and IoT devices to monitor and predict the patient's mental and physical condition. Managing health records, insurance records, pharmacy prescriptions, etc. makes the system even more complex. Also, with the increase in the number of aged and chronically ill patients, the environmental and clinical setup required for monitoring patients and providing an assistive environment is challenging and difficult.

4.3.6 Reliability

Reliability is one of the factors for determining the quality of healthcare. When the network infrastructure is unreliable, the efficiency of the system degrades [24]. If the health environment is not fault-tolerant, then the accuracy and integrity of the data is lost. This prevents the doctor from providing the appropriate treatment and giving real-time assistance to patients who are in need. The healthcare network should have sufficient redundancy of all hardware to maintain robustness. In addition, during technology migration there must be a smooth transition from the legacy setup to the upgrade in terms of both hardware and software. This has to be ensured in order to provide high-quality service.

4.4 Existing technologies

The existing literature was explored to understand the various technologies that have already been adopted in providing healthcare services. Each framework varies in its own way based on applicability of different use-cases. This section elaborates on the existing framework for providing smart healthcare.

4.4.1 Telemedicine

Telemedicine is a computer-based technology that moves beyond boundaries and provides clinical services from a remote location through the telecommunication system without an in-person visit. This technology facilities the real-time exchange of specialty skills and allows collective decisions to be made with remote colleagues.

Patients can be offered treatment by physicians with multiple specialized skills, which would be otherwise impossible because of shortages of skilled physicians. It is a widely used means of treatment for elderly and chronic ill patients residing in remote and rural areas. There is an increase in the usage of telemedicine for adult surgery specialists [25].

This technology can be used in the preparation phase, during treatment, during posttreatment and follow-ups, which contributes to the convenience, timeliness, and cost-effectiveness of the healthcare system. The full potential of telemedicine applications can be realized if the patients and doctors are present in a fixed location where the telemedicine infrastructure is available. But if the mobility of the patients and doctors is problematic, then it can be a challenging task to ensure security and privacy [26]. Managing telemedicine technology comes with added pressures, such as doctor-patient communication, emergency diagnosis, patient's record transmissions, etc.

4.4.2 Smart healthcare

One answer to the rapid rise in the need for medical practitioners to care for chronically ill patients is ubiquitous healthcare, which can be envisioned as a combination of various entities such as on-body sensors, wearable devices, smart hospitals, and information and communication technology (ICT), including the traditional technologies. The smart healthcare system is envisioned to deal with emergency situations and handling them in real time. Elderly patients can be facilitated to have independent living with minimal need for assistance from nurses or caretakers. Smart healthcare enables quality care to be provided where doctors, nurses, treatment, medications, and health-related information are all available for access around the clock [27]. Fig. 4.3 depicts the overview of a smart health care system in which interaction takes place via the cloud.

4.4.3 PDA-based healthcare

The HealthPAL system [28] was proposed to provide an intelligent dialogue-based system to assist elderly patients who lack IT skills. It enables aging and chronic patients to access health-related information daily in an autonomous and independent manner without the assistance of nurses or caretakers. This system stores the patient's medical records, diagnosis report, treatment specification, medicine prescription, symptoms and vital signs, behavioral patterns, etc. A summarized profile of the patient is available as a referential source for both doctor and patients for further treatment and quality care. To address emergency situations, the dialogue recognizes speech and gestures and alerts are issued to notify the doctor via dialed calls or SMS about the urgency in treating the patient. The system is augmented with a medical database to help the elderly user to estimate their health condition and actively decide on how to respond to symptoms. The database contains a pocket

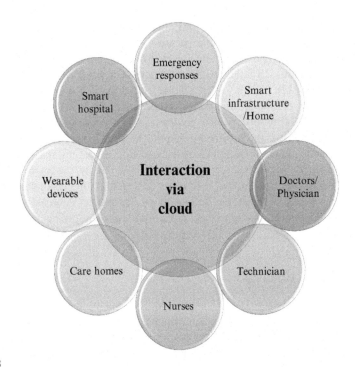

FIG. 4.3

Overview of smart healthcare system.

medical reference book that gives the symptoms and required treatments and a handy medical dictionary that describes the various medical concepts.

4.4.4 Ubiquitous healthcare

In order to care for chronically ill patients, the ubiquitous healthcare system was introduced. The system that was proposed in [9] is a patient-centric system where a variety of sensors such as ECG, EMG, body temperature, and monitoring sensors are embedded into a device that may be attached to the patient directly or indirectly. The data collected are transmitted through a wireless interface to the health information management server for storage and processing. The server content is accessible to the doctor for analysis and offering treatment to the patients. Through smartphone apps, queries can be sent for retrieval of health records of the patient.

Based on the literature review carried out, it has been observed that an intelligent and enhanced system is needed. Due to the critical and heterogeneous nature of the data, an improved secure and reliable healthcare system is proposed to transmit the acquired data across the network. Further, the novelty of this framework is to support practitioners with an infrastructure that provides ubiquitous healthcare with ambient intelligence and cognitive analytics to provide quality care. The following section elaborates on the framework and its components.

4.5 **Proposed framework**

With the increase in the number of patients with chronic diseases, the need for continuous monitoring of the state of health at substantially low cost is becoming vital. Ensuring quality healthcare is the target of most of the e-health applications. One of the possible solutions for reducing healthcare cost and providing timely treatment with proper medication and safety is through ubiquitous healthcare technology, which uses a large number of sensors, wearables, and ubiquitous devices to monitor patients. These devices generate pools of healthcare-related data from electronic health records, genomic data, clinical factors, lab results, pathology reports, MRI images, CT scans, X-rays, and unstructured texts, apart from exogenous sources from wearable devices. These meaningful data are exploding in terms of volume, variety, velocity, and veracity. This technology would enable moving from a doctor/hospital-centric model to a patient-centric healthcare service, apart from supporting the new paradigm shift of medical services in relation to diagnosis and treatment by providing supplementary information to medical doctors.

However, there is a need for technology that can handle such rich heterogeneous resources to monitor the patient's status in real time. Cognitive computing is an emerging trend to merge human and machine potential. The benefit of this trend is being exploited in the healthcare sector in assisting healthcare practitioners in decision making. The ultimate goal of cognitive computing in healthcare is to generate an automated information technology system that is capable of solving problems without human intervention through self-learning techniques. Cognitive techniques such as data mining, machine learning, natural language processing, and others will generate predictive models by analyzing the heterogeneous dataset and provide a cost-effective diagnosis with more certainty.

This paper proposes a novel and integrated ubiquitous healthcare framework that would work to assist chronically impaired patients. In this framework, we introduce several technologies that cooperatively work together to analyze the health state and provide emergency alerts to the practitioners in real time. The framework overcomes the limitations that are faced by the traditional healthcare system. Fig. 4.4 shows a detailed illustration of the proposed framework supported by UC, cognitive computing, and AmI by leveraging the benefits of edge, fog, and cloud computing. The framework consists of four tiers, namely ubiquitous devices, edge computing, fog computing, and cloud computing with big data and cognitive analytics.

4.5.1 **Tier 1: Ubiquitous devices**

The first level represents the ubiquitous devices that are part of the system. The devices are categorized into four types. The first category represents sensors, gadgets, activity recognition devices, and other end devices. Since the goal is to prevent risks by providing medications at anytime and anywhere, it is important to use many environmental biosensors and devices to monitor the patients. Data collected from the environmental sensors, biosignal sensors, and brain-computer interface signals from the patients are transferred to the user terminal to continuously monitor and

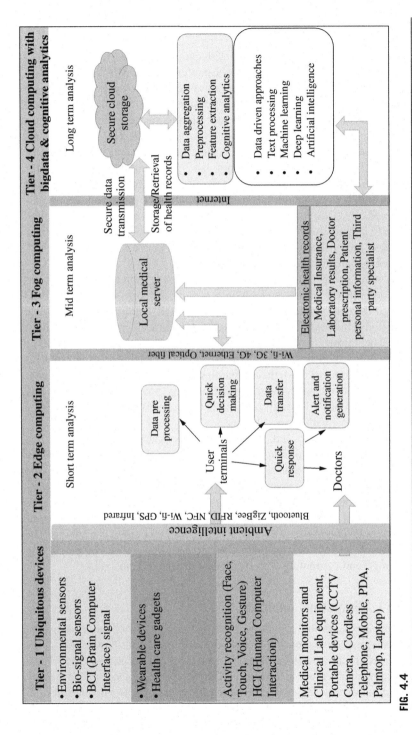

FIG. 4.4

UC healthcare framework.

raise quick alerts. A biosignal, also called a bioelectric signal, is measured and monitored from the human body. The sensors can be subcategorized for measuring electroencephalograms (EEG), heart rate from electrocardiograms (ECG), respiration electromyogram (EMG), electrooculogram (EOG), finger temperature (FT), carbon dioxide (CO_2), and oxygen saturation (SpO_2). In some cases, chronic patients with brain injuries will be in a persistent vegetative state or a minimally conscious state. Brain signal acquisition will help to understand the health state and response to drugs.

The revolution in the miniaturization of electronic devices has facilitated the development of a body area network (BAN). This approach has resulted in improved health detection by continuously monitoring health features such as temperature, pressure, heart rate, etc. The devices that are part of this network are wirelessly connected. The devices are mounted on the body surface, in clothing, or implanted inside the body. This enables response to emergency situations and care for patients in real time.

The second category includes gadgets and wearable devices such as a tracker watch, brain-sensing headband, wristband, breath analyzer, pain relief device, smart walking stick, smart fork, smart clothing, bioscarf, and other health devices. They are used to continuously monitor and study the health status of an individual for vital signs and behavior and assist in the caring for the patients by notifying caretakers via alerts and notifications.

The third category involves human activity recognition wearable devices, which are essential for some chronic patients who are unable to express their needs. The gestures and movements of chronic patients are acquired using wearable devices.

The introduction of human-computer interaction in healthcare is another breakthrough in the field of UC. The availability of tablets, smartphones, tracking devices, depth cameras, wearable devices, augmented reality goggles, Google Glass, Leap Motion, Eyegaze, etc. has increased the opportunities for applying this technology in the UC domain. This technology is beneficial for chronic patients, especially for paralytic patients, to express their needs by accessing the devices through brain signals. The acquired signal is transmitted to the computer and translated into a command or characters on an onscreen device or voice.

The fourth category represents different portable devices such as laptops, PDAs, palmtops, cordless phones, smartphones, medical monitors, and clinical equipment. All these categories of ubiquitous devices act as a user interface between the patient and the user terminal that gathers all the information for medical analysis and diagnosis. The data collected from the UC devices are transferred to the ambient intelligence layer and to the tier-2 terminals for further processing.

Beneath the tier-1 functionalities lies the ambient intelligence layer. Ambient intelligence facilitates the integration of sensor details and provides intelligent decisions that are sensitive and responsive to the chronic patient's needs. It successfully interprets the contextual information from the heterogeneous sensors and meets the need in a timely and transparent manner. It can anticipate the needs and adapt to the user's environment.

As UC relies heavily on data storage and real-time computation, the framework has been designed as a 4-tier architecture with edge, fog, and cloud storage and computing infrastructures to provide time-bound solutions. Additionally, it will provide a variety of computing and data storage resources at different layers.

4.5.2 Tier 2: Edge computing

This level of the framework incorporates edge computing. It allows processing to be performed locally at multiple decision points for the purpose of reducing network traffic. The compute and storage systems residing at the edge layer are very close to the component, device, application, or human that produces the data. The processing latency is very low as the data need not be sent from the edge of the network to a central processing system to be analyzed, and vice versa.

Here all the captured data are received and preprocessed by the user terminal for taking immediate action. The data are processed within the local network before they are sent to the cloud for further processing and analysis. The preliminary decisions are made at the device or at the user terminal to generate alerts or notifications to inform the patient's caretakers about emergency situations so that assistance can be offered to the patient.

4.5.3 Tier 3: Fog computing

The fog layer is the perfect intermediate layer with enough compute, storage, and networking resources to mimic cloud capabilities and to support a quick turnaround of results. Here all the data acquired from the UC devices are transferred from the edge layer and stored at the servers in the fog layer along with the EHRs obtained from laboratories, doctor's prescriptions, discharge summaries, patient personal information, expert opinions, details of insurance claims, etc. Further analysis is carried out with historical and real-time data on the patient's health state using advanced cognitive techniques to notify the practitioners.

4.5.4 Tier 4: Cloud computing with big data and cognitive analytics

As the real-time data acquired are enormous, it is hard to manage and provide a cost-effective storage system. To improve the convenience and security with on-demand access, the health information is stored in the public cloud infrastructure. The confidentiality and privacy of the patient's records and other health-related information can be ensured via encryption techniques. To avoid unauthorized access of the data stored in the cloud, public and symmetric key-based encryption schemes are used. This protects from data theft and data access by untrusted parties. This secured data is stored in the cloud for long-term analysis. Thus, the critical data are accessed only by genuine recipients.

The acquired and transmitted data is processed at the server. Data aggregation, preprocessing, feature extraction, and analytics module constitute the computation engine at the cloud. As the first step, the heterogeneous data from different sources are aggregated. Then, the data is preprocessed, normalized, transformed, and discretized. These raw data are integrated to infer user behavior and health conditions. In the feature extraction step, related techniques are used to extract the representative information for extended processing. In the analytics module, advanced machine learning techniques and text mining algorithms are used to detect and predict anomalies from the integrated heterogeneous data. This framework is suitable to predict the trend of a patient's health over time by comparing the historical data of other patients with similar conditions. The results are exhibited in terms of graphical representations and textual messages to digital electronic devices.

Through cognitive analytics, practitioners can understand the patient condition and decide on the patient-specific treatments. The predictive analytics will also reduce readmissions and ensure the quality of treatment. In addition, it is used to analyze the large volume of medical texts such as literature, scientific discoveries of new relationships between diseases, proteins, genes, drugs and chemicals. It further ensures in generating new perceptions for future research and in developing new tools, frameworks, and techniques.

4.6 Conclusion

Healthcare systems have evolved over the years to provide quality and timely treatment. However, with the new emerging technologies, the existing healthcare system can be improved to support scalability, robustness, integrity, heterogeneity, and security. With the proposed UC health framework, heterogeneous data from multiple sensors, gadgets, medical devices, etc., are acquired from chronic patients by UC devices. A 4-tier storage and computing architecture is proposed to exploit the features of newly emerging technologies such as edge, fog, and cloud infrastructures with cognitive analytics. In each of the tiers, the heterogeneous data is analyzed using advanced cognitive techniques to provide short-term, mid-term, and long-term health status. In order to preserve the privacy and confidentiality of the patient's data, security principles are applied. The 4-tier architecture reduces the round-trip delay in accessing and analyzing the data, which in turn avoids the consequences of delayed treatments.

References

[1] J. Bardram, Hospitals of the future–ubiquitous computing support for medical work in hospitals, in: 2nd International Workshop on Ubiquitous Computing, 2003.

[2] S. Kim, S. Kim, User preference for an IoT healthcare application for lifestyle disease management, Telecommun. Policy 42 (4) (2018) 304–314.

[3] N. Archer, K. Keshavjee, C. Demers, R. Lee, Online self-management interventions for chronically ill patients: cognitive impairment and technology issues, Int. J. Med. Inform. 83 (4) (2014) 264–272.

[4] M. Friedewald, O. Raabe, Ubiquitous computing: an overview of technology impacts, Telematics Inform. 28 (2) (2011) 55–65.

[5] A. Ghose, P. Sinha, C. Bhaumik, A. Sinha, A. Agrawal, A. Dutta Choudhury, UbiHeld, in: Proceedings of the 2013 ACM conference on Pervasive and Ubiquitous Computing Adjunct Publication—UbiComp '13 Adjunct, 2013, pp. 1255–1264.

[6] G. Riva, Ambient intelligence in healthcare, Cyberpsychol. Behav. 6 (3) (2003) 295–300.

[7] A. Chibani, Y. Amirat, S. Mohammed, E. Matson, N. Hagita, M. Barreto, Ubiquitous robotics: recent challenges and future trends, Robot. Auton. Syst. 61 (11) (2013) 1162–1172.

[8] S.F. Ochoa, D. López-De-Ipiña, Distributed solutions for ubiquitous computing and ambient intelligence, Futur. Gener. Comput. Syst. 34 (2014) 94–96.

[9] S. Kim, S. Yeom, O.J. Kwon, D. Shin, D. Shin, Ubiquitous healthcare system for analysis of chronic patients' biological and lifelog data, IEEE Access 6 (2018) 8909–8915.

[10] G. Acampora, D.J. Cook, P. Rashidi, A.V. Vasilakos, A survey on ambient intelligence in health care, Proc. IEEE Inst. Electr. Electron. Eng. 101 (12) (2013) 2470–2494.

[11] D. Oosterlinck, D.F. Benoit, P. Baecke, N. Van de Weghe, Bluetooth tracking of humans in an indoor environment: an application to shopping mall visits, Appl. Geogr. 78 (2017) 55–65.

[12] R. Challoo, A. Oladeinde, N. Yilmazer, S. Ozcelik, L. Challoo, An overview and assessment of wireless technologies and coexistence of ZigBee, bluetooth and wi-fi devices, Procedia Comput. Sci. 12 (2012) 386–391.

[13] M. Bakula, P. Pelgrims, R. Puers, Wireless powering and communication for implants, based on a Royer oscillator with radio and near-field links, Sensors Actuators A Phys. 250 (2016) 273–280.

[14] U. Lee, J.-S. Park, M.Y. Sanadidi, M. Gerla, Flow based dynamic load balancing for passive network monitoring, in: Proceedings of the Third IASTED International Conference on Communications and Computer Networks, CCN 2005, 2005.

[15] R. Rahimi, et al., Laser-enabled fabrication of flexible and transparent pH sensor with near-field communication for in-situ monitoring of wound infection, Sens. Actuators B 267 (2018) 198–207.

[16] S. Kekade, et al., The usefulness and actual use of wearable devices among the elderly population, Comput. Methods Programs Biomed. 153 (2018) 137–159.

[17] H. Tarakci, S. Kulkarni, Z.D. Ozdemir, The impact of wearable devices and performance payments on health outcomes, Int. J. Prod. Econ. 200 (2018) 291–301.

[18] Y. YIN, Y. Zeng, X. Chen, Y. Fan, The internet of things in healthcare: an overview, J. Ind. Inf. Integr. 1 (2016) 3–13.

[19] R. Khan, Future Internet: the internet of things architecture, possible applications and key challenges, in: 2012 10th International Conference on Frontiers of Information Technology (FIT), 2012, , pp. 257–260.

[20] M. Chen, W. Li, Y. Hao, Y. Qian, I. Humar, Edge cognitive computing based smart healthcare system, Futur. Gener. Comput. Syst. 86 (2018) 403–411.

[21] A. Anouar, T. Abdellah, T. Abderahim, A novel reference model for ambient assisted living systems' architectures. Appl. Comput. Inform. (2018), https://doi.org/10.1016/j.aci.2018.08.005.

[22] H. Mshali, T. Lemlouma, M. Moloney, D. Magoni, A survey on health monitoring systems for health smart homes, Int. J. Ind. Ergon. 66 (2018) 26–56.

[23] P. Sarkar, D. Sinha, Application on pervasive computing in healthcare—a review, Indian J. Sci. Technol. 10 (3) (2017) 1–10.

[24] M. Rawashdeh, M.G.A.L. Zamil, M.S. Hossain, S. Samarah, S.U. Amin, G. Muhammad, Reliable service delivery in tele-health care systems, J. Netw. Comput. Appl. 115 (2018) 86–93.

[25] M.T. Harting, et al., Telemedicine in pediatric surgery, J. Pediatr. Surg. 54 (2018) 587–594.

[26] F. Rezaeibagha, Y. Mu, Practical and secure telemedicine systems for user mobility, J. Biomed. Inform. 78 (2018) 24–32.

[27] S.P. Mohanty, U. Choppali, E. Kougianos, Everything you wanted to know about smart cities: the internet of things is the backbone, IEEE Consum. Electron. Mag. 5 (3) (2016) 60–70.

[28] A. Komninos, S. Stamou, HealthPal: an intelligent personal medical assistant for supporting the self-monitoring of healthcare in the ageing society, in: Proceedings of Ubi-Health 2006, 2006.

Dynamic and static system modeling with simulation of an eco-friendly smart lighting system

5

Titus Issac, Salaja Silas, Elijah Blessing Rajsingh

Karunya Institute of Technology and Sciences, Coimbatore, India

5.1 Introduction

In recent years, advances in nanofabrication and embedding technologies have led to smaller, smarter, and more energy efficient devices. A smart device is able to sense, process, control, and communicate dynamically as per the application requirements. Incorporation of smart devices in our everyday lives has led to the modern era of smart computing.

A smart city has digitized its core services and integrates smart devices for these services [1–4], with a controller connected through a wired or wireless medium to remotely monitor and control the smart devices. After the advent of IPV6 in the Internet of Things (IoT), every smart device could be (i) identified by a unique IP address, (ii) monitored, and (iii) controlled from any part of the world [5].

Illumination, parking, water, and traffic management are a few notable core services of a smart city [4]. However, illuminating a city consumes a major fraction of global power supplies, thus escalating the need to investigate existing lighting systems [1, 2, 6, 7]. A lighting system is composed of a set of luminaires. Primarily, a luminaire contains one or more lighting units. A block diagram of a smart luminaire is depicted in Fig. 5.1.

The auxiliary units of a luminaire contain sensors, communication, and processing units. A conventional luminaire [8] is mounted with a lighting unit, whereas a smart luminaire contains sophisticated lamp control drivers [9], sensors [10], and communication modules [4]. A lamp control driver enables the luminaire to achieve the desired luminous levels [9]. Modern generation luminaires are embedded or empowered with multiple sensors, such as a proximity sensor, motion detection sensor, luminous sensor, occupancy sensor, etc. [9, 11], and communication devices, paving the way to a new generation of smart lighting systems [4, 12]. Existing legacy luminaires can also be transformed into smart luminaires with modern retrofits [3, 13].

Systems Simulation and Modeling for Cloud Computing and Big Data Applications
https://doi.org/10.1016/B978-0-12-819779-0.00005-8

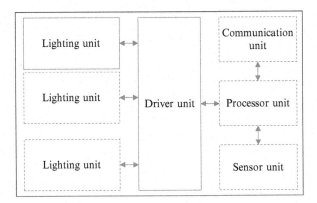

FIG. 5.1

Block diagram of a smart luminaire.

A lighting system can be broadly classified as an (i) outdoor or (ii) indoor lighting system. The outdoor lighting systems include street lighting [10] and traffic control [7]. The indoor lighting systems include home lighting [10], office lighting [6], supplementary lighting systems for plants [14], etc. Compared to an indoor lighting system, the outdoor street lighting system in a city involves (i) a higher quantity of luminaires; (ii) higher cost due to procurement, installation, maintenance, and power; and (iii) high power consumption. A street lighting system requires a set of luminaires with high illumination intensity. The level of the illumination intensity (L) is directly proportional to the power consumption (P) and cost of the luminaire [15, 16].

A city is considered to be safer when adequately illuminated [13]. The level of street illumination influences the rate of crime, fear, and pedestrian movement [13]. Safety illumination measures to illuminate a city appropriately are a major issue. Factors such as the type of luminaire to use, most effective active luminaire count, and best illumination levels escalate the need for further investigations on street-lighting applications using various kinds of luminaires. However, investigating by procuring different types of luminaires for a smart city scenario would incur huge costs and real-time street illumination testing would not be ideal.

In this chapter, the various factors influencing the lighting systems are enumerated. A smart lighting system can be modeled as a multicriteria decision-making problem by considering these various factors, including luminaires, city zones, and lighting duration. A Centralized Eco-Friendly Role Assignment Scheme (CEFRAS) is proposed and a study of static and dynamic scenarios was performed. Various analyses for the luminaire's active duration (AD), power utilization, and lifetime were carried out based on (i) conventional method, (ii) static, and (iii) dynamic role assignment schemes.

The existing smart lighting systems are reviewed and summarized in the following section on related work. The lighting system for a smart city is modeled in

Section 5.3. The various factors influencing a street-lighting system are enumerated and investigations of CEFRAS in a static and dynamic environment are discussed in Section 5.4. The simulation setup and simulation results are presented in Section 5.5, and the work concludes with the future directions in Section 5.6.

5.2 Related work

Recent trends in lighting systems have been widely discussed in the literature. Various techniques for effectively utilizing modern-day technologies like sensors, communication media, and new types of luminaires in the current scenario are studied. Zenulla et al. [3] proposed an urban IoT architecture with the required devices for deploying various applications in smart cities. A lighting system was visualized by optimizing the intensity of a streetlight, depending on the climatic conditions and detection of people.

Marco et al. [2] envisioned smart street lighting that offers additional functions such as air pollution control and WiFi connectivity through the luminaire. Guillemin et al. [10] introduced an innovative, integrated controller for indoor lighting that controls (i) heating/cooling system, (ii) ventilation, (iii) blinds on windows, and (iv) lighting based on the presence of the user. Outdoor illumination requirements were not considered in the architecture.

Viani et al. [15] utilized a particle swarm optimization-based evolutionary algorithm for smart illumination of artifacts in a museum. The approach employed wireless sensor and actuator units to dynamically illuminate the museum on demand. WiFi-based dimmable luminaires were employed for showcasing the artifacts in the museum. The approach was suitable only for indoor scenarios; a large-scale outdoor scenario was not addressed due to the inherent design of the system.

Tan et al. [8] proposed a smart DC grid-powered LED lighting system using sensors for an indoor scenario. DC powered LED luminaires achieves a higher lifetime and efficiency with lower maintenance compared with conventional luminaires. Challenges in outdoor illumination were not addressed. Atici et al. [16] used power-line communication (PLC) between luminaires to dynamically adjust luminous intensity. The drawbacks of the PLC were highlighted. Modern-day luminaire controls were not adopted in the work.

Kaleem et al. [11] substantiated the overall advantages of using an LED luminaire over conventional luminaires. A Zigbee outdoor light monitoring and control (ZB-OLC) system was proposed and employed on an LED luminaire to achieve better power efficiency. Large-scale implementation of the proposal was not addressed. Su et al. [4] proposed the realization of smart lighting control in a smart home with various sensor devices. Dynamic control of illumination and outdoor illumination measures were not considered. Juan et al. [17] proposed a predictive centralized architecture for a smart lighting system using an artificial neural network with the ANOVA statistical method. ANOVA has been used to predict traffic and counts of people to generate a suitable lighting schedule. The architecture achieved lower energy consumption in comparison to a legacy approach. Dynamic changes in requirements for illumination were not considered in the system.

In our previous work, we proposed a luminaire-aware centralized outdoor illumination role assignment scheme (COIRAS) [18]. It was a multiobjective, weighted average-based static role assignment scheme. It considered the luminaire properties and a few factors, including illumination intensity, power consumption, lifespan, total illumination duration, temperature, location, and duration, that affect the lighting system. The static role assignment scheme was unable to address the dynamic needs of the lighting system.

In summary, the luminaires were assigned to illuminate at various levels statically without considering the heterogeneity of the luminaires, the zones, and the duration. Emphasis on safety illumination measures without including the illumination level and energy requirements has to be explored more thoroughly. Also, most of the lighting systems in the literature were tested for a small-scale environment. The modeling and simulation of a smart street-lighting system would help to visualize (i) the overall cost, (ii) power utilization, and (iii) lifetime analysis.

5.3 Modeling a smart street-lighting system

The required quantity, cost, and illumination intensity of a street luminaire in a street-lighting system are significantly higher than for indoor luminaires. The street illumination in a city contains different types of luminaires. Procurement and installation of luminaires in real time incurs a huge cost. Small-scale testing may not be appropriate for a very large city. Therefore, system modeling of smart-lighting systems allows various experimentations on the luminaires to be carried out on a very large scale without incurring physical cost, power consumption, or endangerment of people. The following section identifies the major factors influencing the lighting application and models the luminaire, zone, and duration of a lighting system.

5.3.1 Factors influencing smart lighting

A brief discussion of the major factors affecting the lighting system is presented in the following.

(i) **Illumination intensity** [18]: Total luminance emitted by a luminaire. The luminance is measured in lumens.

(ii) **Power consumption** [18]: Total power consumption of the luminaire, including the power consumed by the lighting, sensors, communication, processor, and control components.

(iii) **Lifespan** [18]: The maximum lifetime of a luminaire is the maximum illumination duration until the luminaire illumination intensity level drops below the critical lumen.

(iv) **Total illumination duration** [18]: Total illumination duration is the maximum active duration of a luminaire.

(v) Temperature [18]: High operating temperature reduces the lifetime of a luminaire. The heat sensors mounted on a luminaire are able to detect and monitor heat changes.

(vi) Location [18]: The placement of a luminaire can be categorized into various zones based on its significance. The zones are classified into critical zones (CZs) and noncritical zones (NCZs).

(vii) Duration [18]: Illumination based on duration help to conserve energy, as the luminaire is not required to illuminate continuously. The duration is classified as peak hours (pH) and nonpeak hours (NPH).

5.3.2 System modeling

System modeling helps to understand the inherent functioning of the system. The luminaire, (ii) zone, and (iii) duration modeling are discussed in the following sections.

5.3.2.1 Luminaire modeling

The lighting system comprises a set of heterogeneous luminaires. A city would contain a wide range of heterogeneous luminaires. Fig. 5.2 demonstrates single- and multi-lighting units in a luminaire across various street-lighting scenarios. Fig. 5.2A depicts

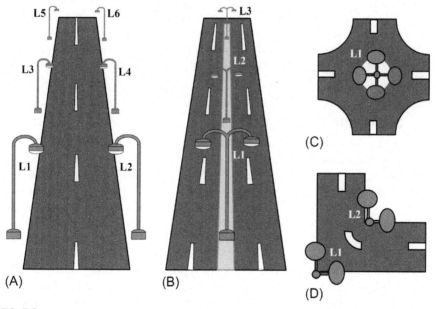

FIG. 5.2

Various kinds of existing single and multiluminaires for street lighting. (A) Street lighting; (B) street lighting multiple roads; (C) street junction lighting; (D) street corner lighting.

a set of luminaires with a single lighting unit deployed across a road in a street. Fig. 5.2B–D depicts multiple lighting units in a luminaire deployed in multiple roads, street junctions, and corner scenarios.

5.3.2.2 Zone modeling

The regions in a city are modeled as critical zones (CZs) and noncritical zones (NCZs) [19]. A CZ is a very vital and significant area like underground tunnels, bridges, historic locations, tourist places, bus stops, road junctions, and corners. The number of luminaires and level of luminous intensity across a critical zone are generally higher than in a noncritical zone.

5.3.2.3 Duration modeling

Street illumination with maximum illumination intensity is not required for the entire lighting duration. Based on the frequency of people activity, the duration could be broadly classified as peak and nonpeak hours. Luminaires have to be well illuminated during peak hours and turned off or dimmed during nonpeak hours [17].

The CZ mandates sufficient illumination both in peak and nonpeak hours. Noncritical zones may not require multiple luminaires with maximum illumination level (IL) but may require a prescribed illumination level during the peak hours. The smart lighting system has multiple objectives. They are: (i) to provide sufficient illumination in terms of location and time, (ii) to reduce the power consumption, (iii) to prevent unintended lighting, and (iv) to maximize the luminaire lifetime.

Identifying the major factors in the lighting application and modeling the luminaire, zone, and duration from the perspective of a lighting application can lead to better understanding of an outdoor lighting system. The insights gained are used to effectively assign roles to the luminaires in the following sections.

5.4 Static and dynamic CEFRAS

A CEFRAS, as defined earlier, is proposed for a smart city lighting system. The role assignment is the process of assigning a set of tasks to a device [19, 20]. Primarily, smart lighting systems could have multiple illumination roles based on the requirements. In this work, the following roles are considered: (i) luminaire with full illumination level, (ii) luminaire with average illumination level, and (iii) luminaire with low illumination level. Variation in the level of illumination is achieved by utilizing a dimmer circuit [8]. The roles defined may not be applicable to all kinds of luminaires, especially the legacy luminaires.

The process of role assignment is performed periodically. During the role assignment, the luminaire's temperature and current role are obtained periodically from the luminaire sensing units and are processed by a powerful central controller (CC). Legacy luminaires are retrofitted with wireless sensors and luminaire controllers. The roles are assigned to a luminaire after the evaluation of the parameters using a Technique for Order of Preference by Similarity to Ideal Solution (TOPSIS) method.

5.4.1 **TOPSIS**

TOPSIS is a multicriteria decision analysis method based on the shortest geometric distance from the positive ideal solution and the longest geometric distance from the negative ideal solution [21]. The TOPSIS from the work [21] is adapted for the CEFRAS and is illustrated here.

Step 1. Generate $n \times m$ evaluation matrix **E** where e_{ij} is the matrix element of matrix **E**. The matrix element $e_{i,j}$ is the performance metric of the luminaire L_i ($1 \leq i \leq n$) and criteria C_j ($1 \leq j \leq m$).

Step 2. Calculate the normalized decision matrix **F**, where f_{ij} is the matrix element of matrix **N**. The value of f_{ij} is calculated by Eq. (5.1) where $i = 1, 2, ...n, j = 1, 2, ..., m$.

$$f_{ij} = e_{ij} / \sqrt{\sum_{j=1}^{m} (e_{ij})^2} \tag{5.1}$$

Step 3. Calculate the weighted normalized decision matrix **G** where v_{ij} is the matrix element of matrix **G**. The matrix is obtained by using Eq. (5.2) where **W** is the criteria weight matrix containing w_j elements and $\sum_{j=1}^{m} w_j = 1, j = 1, 2, ..., m$.

$$G = F \times W \tag{5.2}$$

Step 4. Calculate the best luminaire (L_b) and worst luminaire (L_w) using Eqs. (5.3), (5.4) where J_-, J_+ are the set of j criteria with negative and positive impacts.

$$L_b = \{ \langle \min (v_{ij} \mid i = 1, 2, ..., n) \mid j \in J_- \rangle, \langle \max (v_{ij} \mid i = 1, 2, ..., n) \mid j \in J_+ \rangle \} \tag{5.3}$$

$$L_w = \{ \langle \max (v_{ij} \mid i = 1, 2, ..., n) \mid j \in J_- \rangle, \langle \min (v_{ij} \mid i = 1, 2, ..., n) \mid j \in J_+ \rangle \} \tag{5.4}$$

Step 5. Calculate the L2-distance between the best luminaire L_b and the worst luminaire L_w where d_{iw} and d_{ib} are the distances from the luminaire l to the worst and best conditions, respectively.

$$d_{iw} = \sqrt{\sum_{j=1}^{n} (v_{ij} - v_{wj})^2}, i = 1, 2, ..., n \tag{5.5}$$

$$d_{ib} = \sqrt{\sum_{j=1}^{n} (v_{ij} - v_{bj})^2}, i = 1, 2, ..., n \tag{5.6}$$

Step 6. Calculate the relative closeness to the ideal solution.

$$LFV_i = \frac{d_{iw}}{d_{ib} + d_{iw}}, i = 1, 2, ..., n \tag{5.7}$$

At the end of the TOPSIS method, the luminaire fitness value $(LFV)(0 \leq LFV \leq 1)$ is calculated for every luminaire. If the LFV is 1 or 0, the luminaire has the best or worst solution, respectively.

The luminaires perform the assigned illumination task corresponding to their role. The various phases of static and dynamic CEFRAS are discussed in the following.

5.4.2 Static-CEFRAS

The overview of phases of static-CEFRAS is depicted in Fig. 5.3.

   (i) *Registration phase:* Initially, all the *n* luminaires (L) in the city are registered with the CC with their luminaire properties and location using *LUMINAIRE_INFO message.*

  (ii) *Initialization phase:* The central controller generates a *ROLE_ASSIGNMENT_INIT* message with a specific timeout period and broadcasts it to all the luminaires. The luminaires acknowledge the init message with a property update message *ROLE_ASSIGNMENT_INIT_ACK$_i$* containing the current luminaire role, total power consumption, and illumination duration.

 (iii) *Evaluation phase:* After the *ROLE_ASSIGNMENT_INIT timeout* period, every luminaire is evaluated and assigned a luminaire fitness value (LFV) based on

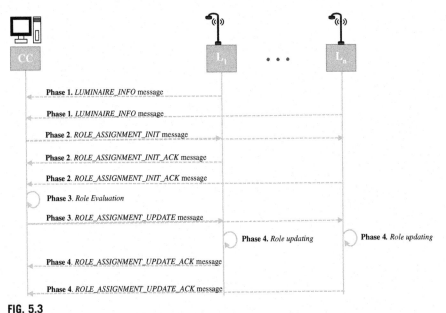

FIG. 5.3

Overview of static-CEFRAS.

the TOPSIS algorithm for m criteria values (c_k) $(1 \leq k \leq m)$ and the corresponding criteria weights (w_k) in the CC. The roles are assigned based on the LFV. The role assignment of all the luminaires is sent via the ROLE_ASSIGNMENT_UPDATE message.

(iv) *Reorganization phase:* The luminaire receives the ROLE_ASSIGNMENT_UPDATE and assigns the role specified in the update message. After updating, an acknowledgement message ROLE_ASSIGNMENT _UPDATE _ACK is dispatched to the CC. The pseudocode for the static-CEFRAS is provided in Algorithm 5.1.

Algorithm 5.1: Static-CEFRAS

```
 1: for all i ∈ Lₙ do
 2:        send LUMINAIRE_INFO to CC
 3: end for
 4: send ROLE_ASSIGNMENT_INIT to 'n' luminaires
 5: while ROLE_ASSIGNMENT_INIT.expiry = false do
 6:        for all i ∈ Lₙ do
 7:               send ROLE_ASSIGNMENT_INIT_ACKᵢ to CC
 8:        end for
 9: end while
10: SLFV=TOPSIS
11: Assign role to Nᵢ based on SLFV; i ∈ 1 ≤ n
12: Send ROLE_ASSIGNMENT_UPDATE to all L
13: for all i ∈ Lₙ do
14:        send ROLE_ASSIGNMENT_UPDATE-ACK to CC
15: end for
```

5.4.3 Dynamic-CEFRAS

The phases in the dynamic-CEFRAS are (i) registration phase, (ii) initialization phase, (iii) evaluation phase, and (iv) reorganization phase. The overall phases of the dynamic-CEFRAS has been depicted in Fig. 5.4. Every luminaire is preprogrammed with a set of static roles in the beginning.

(i) *Registration phase:* An unregistered Luminaire (L_i) $(1 \leq i \leq n)$ performs a one-time registration with the CC using a $LUMINAIRE\_INFO_i$ message upon joining the system.

(ii) *Initialization phase:* The luminaire dynamically generates a role change ROLE_ASSIGNMENT_INIT message based on the (i) luminaire's current role, (ii) active duration, (iii) temperature, (iv) traffic and people (motion detector, proximity sensor), and (v) weather (lunar cycle).

(iii) *Evaluation phase:* Upon receiving the ROLE_ASSIGNMENT_INIT message, every luminaire is evaluated and assigned an LFV based on the TOPSIS algorithm for m criteria values (c_k) $(1 \leq k \leq m)$ and the corresponding criteria

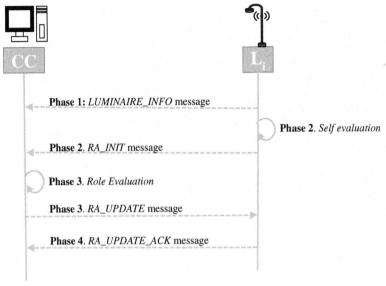

FIG. 5.4

Dynamic-CEFRAS.

weights (w_k) in the CC. The role assignment for all the luminaires is sent via the *ROLE_ ASSIGNMENT_UPDATE* message.

(iv) *Reorganization phase:* The luminaire (Li) receives the *ROLE_ASSIGNMENT_UPDATE* and updates itself with the specified role using the update message. An acknowledgement message *ROLE_ ASSIGNMENT_UPDATE_ACK$_i$* is sent to the CC for the acceptance of the role be L$_i$. The pseudocode for the dynamic-CEFRAS is provided as Algorithm 5.2.

Algorithm 5.2: Dynamic-CEFRAS

```
1: if Lj. status = unregistered then
2:         send LUMINAIRE_INFOi to CC
3: end if
4: if Lj. ROLE_ASSIGNMENT.Initiate = true
5:         send ROLE_ASSIGNMENT_INIT to CC
6: end if
7: SLFV=Topsis
8: Assign role to Ni based on SLFV; i ∈ 1 ≤ n
9: Send ROLE_ASSIGNMENT_UPDATE
10: for all i ∈ Ln do
11:         send ROLE_ASSIGNMENT_UPDATE_ACK to CC
12: end for
```

The performance analysis of the proposed methods static-CEFRAS and dynamic-CEFRAS is carried out by simulation in the following section.

5.5 **Simulation analysis and results**

Simulation analyses were performed on various kinds of luminaires for smart street-lighting scenarios. The luminaire types used were (i) metal halide (MH), (ii) light emitting diode (LED), and (iii) light emitting diode with driver (LED-WD). TOL1, TOL2, TOL3, TOL4 types of luminaires are legacy based, while TOL5, TOL6 types are smart luminaires having an LED-dimmer (LED-D) unit. The luminaire and its properties are tabulated in Table 5.1.

Each luminaire has its own unique power rating that reflects the output illumination level. TOL5 and TOL6 types of luminaires have different levels of power ratings based on the level of illumination. The simulations were carried out using Scilab software. Fig. 5.5 depicts the randomly positioned luminaires in a range of 10×10 km. A total of 100 luminaires was taken for the simulation study for a set of 6 luminaires. The individual luminaire count (LC) is tabulated in Table 5.2. All the luminaires could communicate with the local controller unit via a Zigbee communication module as demonstrated in [11] and the local controller could communicate with the central controller with communication cables (PLC) as in [16]. The zone in the dotted circles represents the CZ, while the rest of the region is treated as an NCZ.

Investigations were carried out on (i) conventional method (CM), (ii) static-CEFRAS (SRA), and (iii) dynamic-CEFRAS (DRA). The conventional method of illumination fully illuminates all the luminaires regardless of luminaire type, zones, and duration. SRA was performed on an hourly basis on the luminaires, and DRA was performed by preloading a set of roles onto the luminaires. The luminaire triggered the role assignment process upon satisfying various conditions.

Table 5.1 Luminaire properties.

Label	Type of luminaire	Power rating (W)	Efficiency (lm/W)	Lifetime (h) × 1000
TOL1	Metal Halide	150	80	6
TOL2		250		6
TOL3	LED	70	55	30
TOL4		140		30
TOL5	LED-D	70	55	30
		49		40
		35		50
TOL6	LED-D	140	55	30
		98		40
		50		50

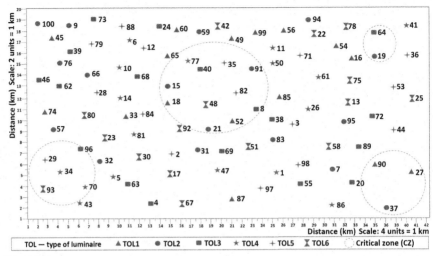

FIG. 5.5

Simulation environment of luminaire application.

Table 5.2 Zonewise luminaire count.

Luminaire label	Type of luminaire	CZ	NCZ
TOL1	Metal Halide	7	4
TOL2	Metal Halide	6	11
TOL3	LED	8	12
TOL4	LED	5	13
TOL5	LED-D	4	11
TOL6	LED-D	5	14

5.5.1 Active duration analysis

The active duration analysis is the total illumination duration analysis of the luminaires. Fig. 5.6 depicts the average active duration of the luminaires for a day. Figs. 5.7 and 5.8 depict the average active duration of the luminaires in the CZ and NCZ. The analysis reveals the active duration of DRA is comparatively higher and lower in CZ and NCZ, respectively. The dynamic role assignment addresses the illumination need of the city by dynamically assigning the role to increase and decrease illumination based on sensor activities, duration, and zone.

5.5.2 Power utilization analysis

The analyses are performed based on the conditions of the luminaires in the two major zones, the CZ and NCZ. Tables 5.3 and 5.4 tabulate the power utilization of the luminaires for a month in the CZ and NCZ using the role assignment schemes

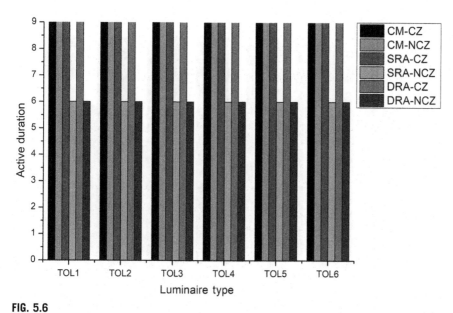

FIG. 5.6

Luminaire active duration analysis.

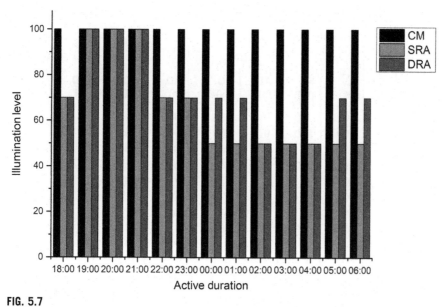

FIG. 5.7

Illumination level during active duration in a critical zone.

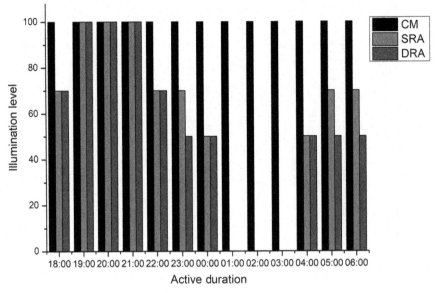

FIG. 5.8

Illumination level during active duration in a noncritical zone.

Table 5.3 Power analysis in critical zone.

TOL	Q	IL (%)	P (W)	CM AD (h)	CM PC (W)	SRA AD (h)	SRA PC (W)	DRA AD (h)	DRA PC (W)
TOL1	7	100	150	9	283.5	9	283.5	9	283.5
TOL2	6	100	250	9	405	9	405	9	405
TOL3	8	100	70	9	151.2	9	151.2	9	151.2
TOL4	5	100	140	9	189	9	189	9	189
TOL5	4	100	70	9	75.6	7	58.8	5	42
		70	49	0	0	2	11.76	1	5.88
		50	35	0	0	0	0	1	4.2
TOL6	5	100	140	9	189	4	84	4	84
		70	98	0	0	3	44.1	2	29.4
		50	50	0	0	2	15	3	22.5

AD, active duration; CM, conventional method; DRA, dynamic role assignment (dynamic-CEFRAS); IL, illumination level; P, power; PC, power consumed; Q, quantity; SRA, static role assignment (static-CEFRAS); TOL, type of luminaire.

Table 5.4 Power analysis in noncritical zone.

TOL	Q	IL (%)	P (W)	CM AD (h)	CM PC (W)	SRA AD (h)	SRA PC (W)	DRA AD (h)	DRA PC (W)
TOL1	4	100	150	9	162	9	162	9	162
TOL2	11	100	250	9	742.5	9	742.5	9	742.5
TOL3	12	100	70	9	226.8	9	226.8	9	226.8
TOL4	13	100	140	9	491.4	9	491.4	9	491.4
TOL5	11	100	70	9	207.9	7	161.7	5	115.5
		70	49	0	0	2	32.34	1	16.17
		50	35	0	0	0	0	1	11.55
TOL6	14	100	140	9	529.2	4	0	4	235.2
		70	98	0	0		164.64	3	123.48
		50	50	0	0	3	63	2	42

AD, active duration; CM, conventional method; DRA, dynamic role assignment (dynamic-CEFRAS); IL, illumination level; P, power; PC, power consumed; Q, quantity; SRA, static role assignment (static-CEFRAS); TOL, type of luminaire.

(i) CM, (ii) SRA, and (iii) DRA. The month of June has the approximate active duration requirement of 9 h and is taken as the maximum active duration [11]. The energy consumed (EC) calculation is performed using Eq. (5.8) and the parameters considered are: (i) power rating (P), (ii) luminaire count (LC), (iii) active duration (D), and (iv) number of days (ND).

$$EC = \frac{P \times LC \times D \times ND}{1000} \tag{5.8}$$

The overall power utilization of the luminaires in the CZ of SRA and DRA in comparison to the CM method is lesser by 3.94% and 5.92%. In the NCZ, the power utilization of SRA and DRA is lesser by 7.31% and 8.18% in comparison to the CM method.

5.5.3 Lifetime analysis

Table 5.5 tabulates the estimated lifetime of the luminaires in days. The lifetime (LT) calculation using Eq. (5.9) is performed based on the luminaire's parameters: (i) estimated lifetime (P), (ii) active duration (D), (iii) total active duration (TD), and (iv) projected active duration of the day (PD).

$$LT_i = \frac{(P_i \times D_i)}{(TD \times PD)} \tag{5.9}$$

The life expectancy of the luminaires remains constant regardless of the CZ and NCZ modes using CM. The life expectancy of the legacy luminaires increased by 33.3% on adopting SRA and DRA. LED-WDs achieve 4.75% more lifetime than LEDs in

Table 5.5 Lifetime analysis.

TOL	P (W)	CM		SRA	DRA	
		CZ (h)	NCZ (h)	NCZ (h)	CZ (h)	NCZ (h)
TOL1	150	667	667	1200	667	1200
TOL2	250	667	667	1200	667	1200
TOL3	70	3333	3333	3704	3333	3704
TOL4	140	3333	3333	3704	3333	3704
TOL5	70	3333	3333	2592	2963	2716
	49	4444	4444			
	35	5556	5556			
TOL6	140	3333	3333	3210	4197	3333
	98	4444	4444			
	50	5556	5556			

the NCZ mode using SRA. The TOL6 type luminaire using DRA achieves 25.94% and 33.34% more lifetime than SRA and CM in the NCZ mode and 20.58% more lifetime than CM in the CZ mode.

5.6 Conclusion

Smart cities are continuously upgrading with smart devices. Smart city lighting systems constituted of modern luminaires with illumination control are found to be much more versatile and energy efficient. System modeling and simulation of various outdoor illumination methods in a smart city were carried out based on factors such as the type of luminaires, duration, and zones. The proposed dynamic-CEFRAS provided better on-demand illumination and achieved a better lifetime with less power consumption than the conventional illumination method and static-CEFRAS. The future direction of this work is to design a role assignment scheme for a distributed environment.

References

[1] C. Boissevain, S. Mcclellan, J.A. Jimenez, Smart cities, in: M.C. Stan, J.A. Jimenez, G. Koutitas (Eds.), Smart Cities Application, Technolgies, Standards and Driving Factors, vol. 16, Springer, 2018, , pp. 181–195. No. 1.

[2] P. Neirotti, A. De Marco, A.C. Cagliano, G. Mangano, F. Scorrano, Current trends in smart city initiatives: some stylised facts, Cities 38 (2014) 25–36.

[3] A. Zanella, N. Bui, A. Castellani, L. Vangelista, M. Zorzi, Internet of things for smart cities, IEEE Internet Things J. 1 (1) (2014) 22–32.

[4] K. Su, J. Li, H. Fu, Smart city and the applications, in: 2011 Int. Conf. Electron. Commun. Control. ICECC 2011—Proc., 2011, pp. 1028–1031.

[5] H. Kopetz, Internet of things, in: Real-Time Systems Design Principles for Distributed Embedded Applications, second ed., Springer, 2011, pp. 307–323.

[6] A. Pandharipande, D. Caicedo, Smart indoor lighting systems with luminaire-based sensing: a review of lighting control approaches, Energy Build. 104 (2015) 369–377.

[7] G. Shahzad, H. Yang, A.W. Ahmad, C. Lee, Energy-efficient intelligent street lighting system using traffic-adaptive control, IEEE Sensors J. 16 (13) (2016) 5397–5405.

[8] Y.K. Tan, T.P. Huynh, Z. Wang, Smart personal sensor network control for energy saving in DC grid powered LED lighting system, IEEE Trans. Smart Grid 4 (2) (2013) 669–676.

[9] I. Chew, V. Kalavally, N.W. Oo, J. Parkkinen, Design of an energy-saving controller for an intelligent LED lighting system, Energy Build. 120 (2016) 1–9.

[10] A. Guillemin, N. Morel, Innovative lighting controller integrated in a self-adaptive building control system, Energy Build. 33 (5) (2001) 477–487.

[11] Z. Kaleem, T.M. Yoon, C. Lee, Energy efficient outdoor light monitoring and control architecture using embedded system, IEEE Embed. Syst. Lett. 8 (1) (2016) 18–21.

[12] P. Neirotti, A. De Marco, A.C. Cagliano, G. Mangano, Politecnico di Torino Porto Institutional Repository, (2014).

[13] C. Basu, et al., Sensor-based predictive modeling for smart lighting in grid-integrated buildings, IEEE Sensors J. 14 (12) (2014) 4216–4229.

[14] Y. Xu, Y. Chang, G. Chen, H. Lin, The research on LED supplementary lighting system for plants, Optik 127 (18) (2016) 7193–7201.

[15] F. Viani, A. Polo, P. Garofalo, N. Anselmi, M. Salucci, E. Giarola, Evolutionary optimization applied to wireless smart lighting in energy-efficient museums, IEEE Sensors J. 17 (5) (2017) 1213–1214.

[16] Ç. Atici, T. Özçelebi, J.J. Lukkien, Exploring user-centered intelligent road lighting design: a road map and future research directions, IEEE Trans. Consum. Electron. 57 (2) (2011) 788–793.

[17] J.F. De Paz, et al., Intelligent system for lighting control in smart cities, Inf. Sci. 372 (2016) 241–255.

[18] T. Issac, S. Silas, E.B. Rajsingh, Luminaire aware centralized outdoor illumination role assignment scheme : a smart city perspective, in: J.D. Peter, A.H. Alavi, B. Javadi (Eds.), Advances in Big Data and Cloud Computing, first ed., Springer, Singapore, 2019.

[19] S. Misra, A. Vaish, Reputation-based role assignment for role-based access control in wireless sensor networks, Comput. Commun. 34 (3) (2011) 281–294.

[20] I. Titus, S. Silas, E.B. Rajsingh, Investigations on task and role assignment protocols in wireless sensor network, J. Theor. Appl. Inf. Technol. 89 (1) (2016) 209–219.

[21] M.S. García-cascales, M.T. Lamata, On rank reversal and TOPSIS method, Math. Comput. Model. 56 (5–6) (2012) 123–132.

Predictive analysis of diabetic women patients using R

R. Rifat Ameena, B. Ashadevi

Department of Computer Science, M.V. Muthiah Government Arts College for Women, Dindigul, India

6.1 Introduction

This work aims at both detecting diabetes as well as predicting the risk of diabetes in a Pima Indian women dataset. The Pima are people of American Indian origin. The framework used here is RStudio using the R programming language. The R framework was selected because it plays a significant role in data analytics and visualization, and RStudio provides a statistical tool with support of machine learning and a visualization language that is easy to learn, provides high code density, is open source and freely available, is easy to install, and provides sophisticated results. It also has huge web support. One of the striking features of RStudio is that it can be combined with Spark and Hadoop, which are required to handle big datasets. Hence, in investigating the causes of many associated health issues like heart attacks, liver failure, kidney failure, and nerve damage, both big data as well as the needed analytics can be handled.

Present-day lifestyles have contributed to the very serious problem of diabetes, a disease involving high blood sugar levels and poor circulation. Women with diabetes have poor immunity, which reduces the body's ability to fight infections. Diabetes is a major factor in many other health problems like heart attacks, obesity, nerve damage, kidney failure, liver failure, high blood pressure, vision loss, and polycystic ovarian syndrome (PCOS). PCOS is occurring more frequently in women nowadays because of the growing trend of high resistance towards insulin. As a result, even teenagers have an increased risk of being diabetic. This condition also causes problems during pregnancy. Hence, diabetes detection and prediction are important concerns in providing better healthcare services, especially for women.

6.1.1 Type 1 diabetes

Type 1 diabetes, in which no insulin is produced by the body, occurs more frequently in children, and is often called juvenile diabetes. In this condition, the insulin-producing pancreatic beta cells are destroyed by the body. Common symptoms are weight loss, dehydration, and damage to body parts such as liver and kidney, vision loss, urinary infections, along with many other issues.

6.1.2 Type 2 diabetes [1]

Type 2 diabetes generally occurs in older people. Both men and women can develop type 2 diabetes, but some symptoms such as urinary tract infection, poly cystic ovarian syndrome, vaginal thrush are more likely to affect women. In Type 2 diabetes, an excess amount of glucose is present in the blood that is left unconsumed. The causes of this type of diabetes are excessive weight, little exercise, hereditary factors, among others. Disorders associated with diabetes are diabetic retinopathy, diabetic neuropathy, malfunctioning of liver, PCOS, stomach paralysis, and infertility. In this type of diabetes, high blood sugar is transferred to unborn babies.

6.1.3 Gestational diabetes

This type of diabetes is specific to pregnant women and it typically occurs later during pregnancy. A glucose tolerance test is used to screen pregnant women for gestational diabetes. In most women, this type of diabetes goes away after pregnancy.

6.2 Literature survey

The dataset used in this study was originally published by the National Institute of Diabetes and Digestive and Kidney Diseases. Accurately diagnosing diabetes in women is very important and, for this purpose, various tools have been built and risk factors identified. However, there is no significant work that can predict or detect diabetes in women with higher accuracy [2–10]. According to surveys conducted, 422 million people have diabetes; by 2035 this may rise to 592 million. The number of people with type 2 diabetes is increasing in every country; 85% of people with diabetes live in underdeveloped and developing countries. The greatest number of people with diabetes are between 40 and 59 years of age, and 175 million people with diabetes are undiagnosed. Diabetes caused 5.1 million deaths in 2013; every 6 s a person dies from diabetes. >21 million live births were affected by diabetes during pregnancy in 2013. In women, pregnancy is one of the supplementary factors, and during pregnancy the mother's excess blood glucose is also transferred to the fetus. As a result, infants of mothers with diabetes are larger than other babies. The excess blood sugar causes the baby's body to produce a higher amount of insulin, which results in increased tissues and fat deposits. Diabetes often strikes women differently when compared to men. This is mainly because of the hormonal variations and inflammation differentiation between the two genders. In consonance with the World Health Organization, women with a 2-h post load plasma glucose level of at least 200 mg/dL have diabetes [11].

Many people have developed prediction models using data mining to predict diabetes [3]. Some of the models developed using data mining are described in this paragraph. Abdulla et al. [1] worked on a predictive analysis of diabetic treatment using a regression-based data mining support vector machine (SVM). Yunsheng et al. used a KNN algorithm by removing the outlier/OOB (out of bag), and in this study the storage space was minimized. After removing a parameter that had a small effect, the researchers got better accuracy. Nilashi et al. [12] used CART (classification and regression tree) analysis, a clustering algorithm (principal component analysis, PCA), and expectation maximization (EM) techniques. They found that some fuzzy rules generated by CART by removing noise were effective in prediction. Velide Phani Kumar et al. [13] analyzed diabetes data using various data-mining techniques such as naïve Bayes, J48 (C4.5) JRip, neural networks, decision trees, KNN, fuzzy logic, and genetic algorithms based on accuracy and time. They found that out of various data-mining techniques J48 (C4.5) took the least time. Rupa Bagdi et al. [2] developed a decision support system that combined the strengths of both online analytical processing (OLAP) and data mining. This system predicted the future state and generated useful information for effective decision making. They also compared the result of the ID3 and C4.5 decision tree algorithms. The system could discover hidden patterns in the data and it also enhanced real-time indicators, discovered obstacles, and improved information visualization. Rajesh et al. [3] carried out research to classify diabetes clinical data and predict the likelihood of a patient being affected with diabetes. The training dataset used for data-mining classification was the Pima Indians Diabetes Database. They applied various classification techniques and found out that the C4.5 classification algorithm was the best algorithm to classify the dataset. Kavakiotis et al. [4] found that machine learning algorithms are very important to predict different medical datasets including diabetes diseases datasets (DDD). In their study, they used SVMs, logistic regression, and naïve Bayes using 10-fold cross validation. The researchers concluded that the SVM algorithm provides the best accuracy. All of these researchers have been successful in analyzing the diabetic dataset and developing good prediction models. But most of them used Weka and Oracle Data Miner data mining tools, and a few of them used tools like Tanagra and Orange. This chapter describes the attempt to analyze the diabetic dataset using R.

6.3 **Proposed work**

As already mentioned, the dataset used for the purpose of this study is the Pima Indians Diabetes Database for women published by the National Institute of Diabetes and Digestive and Kidney Diseases. It contains diagnostic information on women whose age is greater than 20. This database includes 768 females, of which 268 females are diagnosed with diabetes. The samples consist of examples with 8 attribute values and one of two possible outcome values, namely whether the patient has tested positive for diabetes (indicated by output 1) or not (indicated by 0). This dataset has been evaluated using R. Fig. 6.1 shows the architecture of the proposed system.

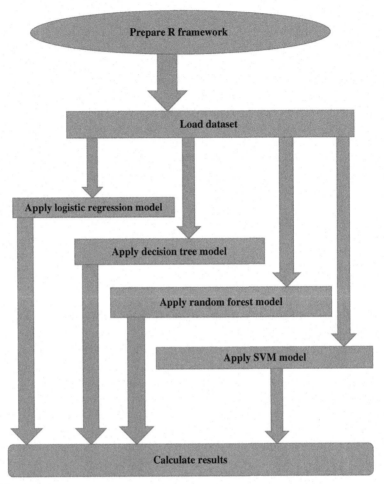

FIG. 6.1

Proposed work.

The diabetic dataset is given as input to the system, by loading into R. The raw data is a comma separated values (CSV) file and appears to be a clutter of data. But a proper evaluation of this dataset reveals some interesting facts. After the raw data is input to R, the dataset is analyzed and partitioned based on different attributes; the output obtained from R is well-formatted data. R is one of the best languages used for statistical computing as well as for generating graphs. Since we know that pictures speak louder than words, after evaluation, graphs were generated for each dataset using R and then the data was plotted. The analysis of the proposed work is shown in Fig. 6.1. The proposed work is discussed in the following section.

6.3.1 **Attributes**

The proposed work makes use of the Pima Indian females diabetes dataset, which is primarily concerned with women's health. Here, 768 instances of women older than 20 years were collected and various parameters defined.

The attributes defined here are:

1. Number of times pregnancy occurred
2. Plasma glucose concentration at 2 h in an oral glucose tolerance test
3. Diastolic blood pressure (mm Hg)
4. Triceps skinfold thickness (mm) 2-h serum insulin (μU/mL)
5. Body mass index (weight in kg/(height in m)^2)
6. Diabetes pedigree function
7. Age (years)
8. Class variable (0 or 1)
   - 0-indicates False diabetic test
   - 1-indicates True diabetic test

6.3.2 **Predictive models**

6.3.2.1 *Logistic regression model*

The logistics regression is common and is a useful regression method for solving the binary classification problem. It is easy to implement and can be used as the baseline for any binary classification problem. Its basic fundamental concepts are also constructive in deep learning. Logistic regression describes and estimates the relationship between one dependent binary variable and independent variables. For predicting diabetes in women, this model is built with the outcome as the response variable, with the rest of the eight variables as predictor variables. Stepwise variable selection is used to find the important variables. In this model, we have used the following code:

```
library(corrplot)
library(caret)
n <- nrow(pima)
pima_training <- pima [train]
```

Implementing the logistic regression model

```
glm_fm1<-glm(Outcome~.,data =pima_train, family=binomial)
step_model<-step(model.glm)
plot(glm_fm2)
lm_probs<- predict (glm_fm2, newdata= pima_testing, type=
"response")
glm_pred <- if else (glm_probs > 0.5, 1, 0)
#print ("Confusion Matrix for logistic regression");
Table (Predicted=glm_pred,
Actual=pima_testing$Diabetes)
confusionMatrix( glm_pred, pima_testing$Diabetes)
```

This model was implemented using RStudio and the result is shown in Fig. 6.2.

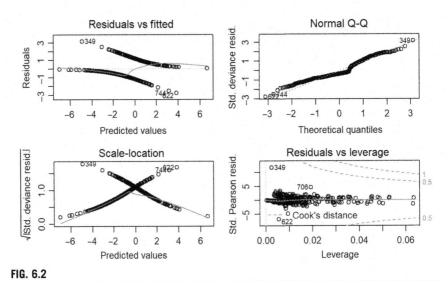

FIG. 6.2

Plot using linear regression model in R.

6.3.2.2 Decision tree

The decision tree is also one of the supervised learning algorithm techniques that is frequently used for prediction. It forms an important decision support tool system and is an integral part of operational research. Hence, the selection of decision tree is an appropriate choice for prediction of diabetes symptoms among women. This is an important step, as early prediction will help in developing the model. When the response variable is categorical, decision tree can be used.

- The Decision Support System is based upon splitting the values based on attributes and conditions, so the graph looks like a tree structure, which can be traversed from root to leaf and the prediction can be evaluated.
- The package used in RStudio for implementing the decision tree is the party package, which has the ctree function used to create and visualize decision trees.
- The syntax used here is ctree (formula, data) where:
- formula: deals with predictor and response variable
- data: is the name of dataset used.

In this dataset, the tree splits the data into two or more homogeneous sets based on the most significant splitter in the predictor value set that is given as input. For diabetes prediction in women using decision tree, we have used the following code:

```
library(tree)
set.seed(1234)
intrain<-createDataPartition(y=pima$Diabetes,p=0.7,list=FALSE)
```

Implementing the decision tree model

```
train <- pima [intrain,] test <- pima [-intrain,]
treemod <- tree (Diabetes ~., data = train) plot(treemod)
text (treemod, pretty = 0)
tree_pred <- predict(treemod, newdata = test, type = "class")
confusionMatrix(tree_pred, test$Diabetes)
```

This model is implemented in R and the result is shown in Fig. 6.3.

6.3.2.3 Random forest model

The random forest is a method of classification that is part of the machine learning model, which combines predictions of weak classifiers. It is widely believed to be the best classifier for high-dimensional data [5]. Random forest works by building a large set of classification/regression trees. Each tree is built with random replacement of the training data. At each split for each tree a random replacement of predictor variables is used. This large set of trees is combined to maximize the performance. It is fast and easy to implement, produces highly accurate predictions, and can handle a very large number of input variables without overfitting. Each tree is formed by selecting at random, at each node, a small group of input coordinates to split and by calculating the best split based on these features in the training set. The tree is grown without pruning. This model for predicting diabetes using R has been implemented using the following code:

```
Set seed (123)
Library (Random Forest)
```

Implementing random forest model

```
rf_probs <- predict (rf_pima, newdata = pima_testing)
rf_pred<-if    else(rf_probs>0.5,1,0)
confusion Matrix (rf_pred, pima_testing$Diabetes)
par (mfrow = c (1, 2))
varImpPlot(rf_pima,type=2,main="Variable Importance",col='black')
plot (rf_pima, main = "Error vs no. of trees grown")
```

This model is implemented in R and the result is shown in Fig. 6.4.

6.3.2.4 Support vector machine

The SVM is one of the supervised learning techniques that are often used for classification of datasets. It aims at forming the hyperplane that is at the maximal distance from the classes during the training phase. During the testing phase, the new instance is calculated on the basis of the maximum distance from the hyperplane so the formed classification can be extended up to multiple class classifications. In that case, formation of the hyperplane is done through kernel formation. SVM is easy to implement and doesn't even require large processing time when dealing with the small dataset of 768 instances. The package e1071

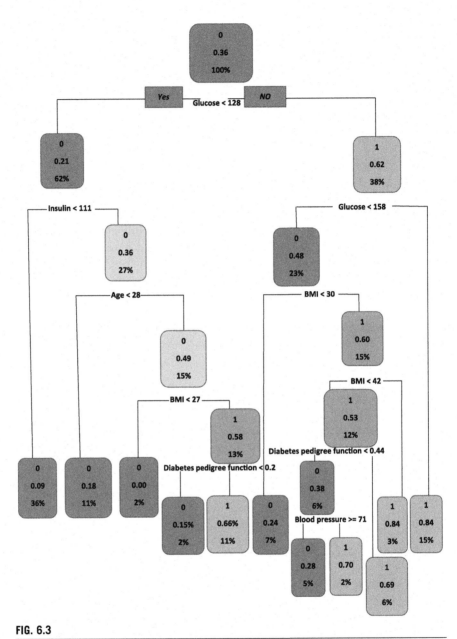

FIG. 6.3

Decision tree generated to predict chances of diabetes in R.

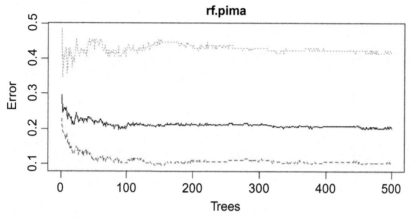

FIG. 6.4

Random forest model generated using R.

was used in RStudio for implementation. The SVM model is implemented using the following code:

```
library(e1071)
pima$Diabetes <-as.factor(pima$Diabetes)
```

Implementing support vector model

```
tuned <- tune. Svm (Diabetes ~., data = train, gamma = 10^ (-6:
-1), cost=10^ (-1:1))
summary(tuned) # to show the results.
svm_pred <- predict(svm_model, newdata = test)
confusionMatrix(svm_pred, test$Diabetes)
```

This model is implemented in R and the result is shown in Fig. 6.5.

Comparing the four models

Comparing the four models, logistic regression, decision tree, random forest, and SVM, we got the following results:

To find the accuracy <-data.frame(Model=c("Logistic Regression", "Decision Tree","Random Forest", "Support Vector Machine (SVM)"), Accuracy=c(acc_glm_fit, acc_treemod,acc_rf_pima,acc_svm_model))ggplot (accuracy,aes(x=Model,y=Accuracy))+geom_bar(stat='identity')+theme_bw () + ggtitle('Comparison of Model Accuracy').

The package e1071 in RStudio was used for implementation. The comparison of accuracy between the predictive models is shown in Fig. 6.6. Among these four predictive models, the random forest model shows the highest accuracy.

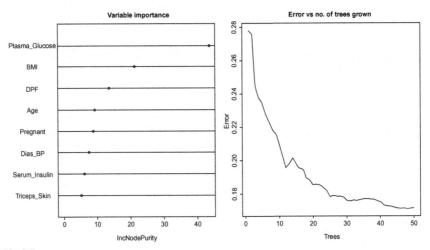

FIG. 6.5

Variable importance with number of trees grown.

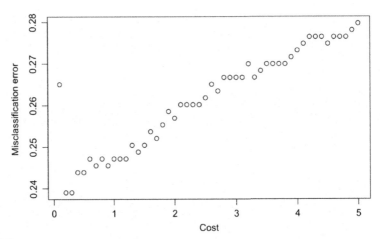

FIG. 6.6

SVM model generated to predict chances of diabetes in R.

6.4 Experimental analysis

The plot in Fig. 6.2 shows that the logistic regression model plots all the points, so there are no outliers. Therefore, the logistic regression model fits perfectly. The error rate is 25.11%, and the accuracy is 74.89%.

Fig. 6.3 shows that there are 5 variables as internal nodes, 14 variables as terminal nodes, and the training error rate is 16.95%. The test rate error is 30% and the accuracy is 70%.

The plot in Fig. 6.4 visualizes the random forest model error. The red line (dark gray in print version) indicates the error rate of predicting if a person does not have diabetes, while the green line (light gray in print version) indicates the error rate of predicting if a person has diabetes. The black line indicates overall error rate. The result produced by this model is a test rate error of 22.04% and the accuracy attained is 77.06%; when the error rate decreases the number of trees grown increases.

Fig. 6.5 plot visualizes important variables for diabetes prediction, which are plasma glucose, body mass index (BMI), and diastolic blood pressure (DPF). This plot depicts that the number of trees grown increases when the error rate decreases.

Fig. 6.6 shows that the cost parameter is tuned to produce the best values that minimize the overall misclassification error rates. Here, the test rate error is 23.04% and the accuracy is 76.96%.

Fig. 6.7 shows the accuracy rate of various prediction models. Among all these models, the random forest model shows the highest accuracy and the decision tree model shows the least accuracy. The accuracy of the prediction model is improved. Thus, the random forest model is more adaptive to the dataset.

Fig. 6.8 shows the matrix of correlation between variables. A correlation matrix is used to summarize data, as an input for more advanced analysis. For this analysis package, corr is used in the R tool. Table 6.1 highlights the classification accuracy of four models namely logistic regression model, decision tree, random forest model, and support vector machine (SVM) we got the following results. Among these models random forest model has the highest accuracy.

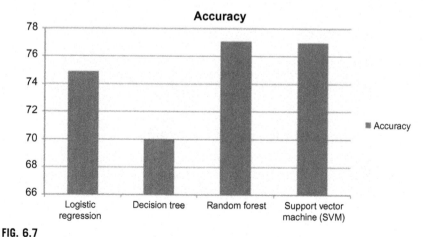

FIG. 6.7

Comparison of model accuracy in R.

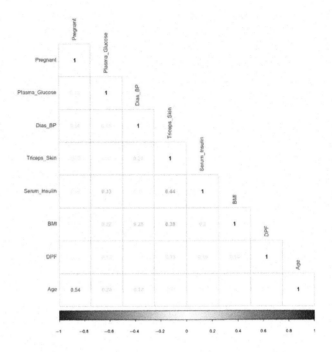

FIG. 6.8

Matrix of correlation between variables.

Table 6.1 Model accuracies.

Models	Accuracy
Logistic regression model	74.89%
Decision tree	70%
Random forest model	**77.06%**
Support vector machine	76.96%

Bold represents highest accuracy of random forest model.

6.5 Conclusion and future enhancements

This paper focuses on analysis of diabetes in women through comparing various prediction models and finding their accuracy with statistical implications using R. We have compared the classification of results using the decision tree, logistic regression model, SVM model, and random forest model. The classification results showed that random forest gives the best results. The random forest with increased classification performance also overcame the overfitting problem generated due to missing values in the datasets. Various data mining techniques and their applications were studied or reviewed. Machine learning algorithms were applied to different medical datasets. Machine learning methods have different functionality in different datasets. The facts revealed during this process can be used for developing some prediction models. The possibilities for future work include an extension study of the different types of techniques to review the feature construction process.

Acknowledgment

We would like to acknowledge all the authors that provided significant help in diabetes research.

References

[1] D. Abdullah, A. Aljumah, M.G. Ahamad, M.K. Siddiqui, Application of data mining: diabetes health care in young and old patients, J. King Saud Univ. Comput. Inf. Sci. 25 (2013) 127–136.

[2] R. Bagdi, P. Patil, Diagnosis of diabetes using OLAP and data mining integration, Int. J. Comput. Sci. Commun. Netw. 2 (3) (2012) 314322.

[3] K. Rajesh, V. Sangeetha, Application of data mining methods and techniques for diabetes diagnosis, Int. J. Eng. Innov. Technol. (IJEIT) 2 (2012).

[4] I. Kavakiotis, O. Tsave, A. Salifoglou, N. Maglaveras, I. Vlahavas, I. Chouvarda, Machine learning and data mining methods in diabetes research, Comput. Struct. Biotechnol. J. 15 (2017) 104–116.

[5] M. Imran, A.M. AlAbdullatif, B.S. AlAwwad, M.M. Alwalmani, S.A. Al-Suhaibani, S. A. Al-Sayah, Towards early detection of diabetic retinopathy using extended fuzzy logic, Int. J. Pharm. Med. Biol. Sci. 5 (2) (2016) 110–114.

[6] R. Meza-Palacios, et al., Development of a fuzzy expert system for the nephropathy control assessment in patients with type 2 diabetes mellitus, Expert Syst. Appl. 72 (2016) 335–343.

[7] Polatkemal, SalihGüne, An expert system approach based on principal component analysis and adaptive neuro-fuzzy inference system to diagnosis of diabetes disease, Digit. Signal 17 (2007) 702–710.

[8] C. Omprakash, S.R. KumarJatinder, Development of Indian weighted diabetic risk score (IW- DRS) using machine learning techniques for type-2 diabetes, in: ACM COMPUTE'16, 21–23 October, 2016.

[9] ButwallMani, kumarShradha, A data mining approach for the diagnosis of diabetes mellitus using random forest classifier, Int. J. Comput. Appl. 120 (8) (2015) 975–8887.

[10] V. Veena Vijayan, C. Anjali, Prediction and diagnosis of diabetes mellitus—a machine learning approach, in: IEEE Recent Advances in Intelligent Computational Systems (RAICS) | 10–12 December 2015 | Trivandrum, 2015.

[11] M. Komi, J. Li, Y. Zhai, X. Zhang, Application of data mining methods in diabetes prediction, in: 2017 2nd International Conference on Image, Vision and Computing (ICIVC), June, IEEE, 2017, pp. 1006–1010.

[12] M. Nilashi, O. bin Ibrahim, H. Ahmadi, L. Shahmoradi, An analytical method for diseases prediction using machine learning techniques, Comput. Chem. Eng. 106 (2017) 212–223.

[13] V.P. Kumar, L. Velide, A data mining approach for prediction and treatment of diabetes disease, Int. J. Sci. Invent. Today 3 (1) (2014).

Further reading

A.A. Al Jarullah, Decision tree discovery for the diagnosis of type II diabetes, in: 2011 International Conference on Innovations in Information Technology (IIT), April, IEEE, 2011, pp. 303–307.

V.R. Balpande, R.D. Wajgi, Prediction and severity estimation of diabetes using data mining technique, in: 2017 International Conference on Innovative Mechanisms for Industry Applications (ICIMIA), February, IEEE, 2017, pp. 576–580.

E.K. Hashi, M.S.U. Zaman, M.R. Hasan, An expert clinical decision support system to predict disease using classification techniques, in: International Conference on Electrical, Computer and Communication Engineering (ECCE), February, IEEE, 2017, pp. 396–400.

S. Kang, P. Kang, T. Ko, S. Cho, S.J. Rhee, K.S. Yu, An efficient and effective ensemble of support vector machines for anti-diabetic drug failure prediction, Expert Syst. Appl. 42 (9) (2015) 4265–4273.

M. Koklu, Y. Unal, Analysis of a population of diabetic patients databases with classifiers, World Acad. Sci. Eng. Technol. Int. J. Med. Health, Pharm.Biomed. Eng. 7 (8) (2013) 481–483.

V.A. Kumari, R. Chitra, Classification of diabetes disease using support vector machine, Int. J. Eng. Res. Appl. 3 (2) (2013) 1797–1801.

S. Mekruksavanich, Medical expert system based ontology for diabetes disease diagnosis, in: 2016 7th IEEE International Conference on Software Engineering and Service Science (ICSESS), August, IEEE, 2016, pp. 383–389.

X.H. Meng, Y.X. Huang, D.P. Rao, Q. Zhang, Q. Liu, Comparison of three data mining models for predicting diabetes or prediabetes by risk factors, Kaohsiung J. Med. Sci. 29 (2) (2013) 93–99.

F. Mercaldo, V. Nardone, A. Santone, Diabetes mellitus affected patients classification and diagnosis through machine learning techniques, Procedia Comput. Sci. 112 (C) (2017) 2519–2528.

A. Pavate, N. Ansari, Risk prediction of disease complications in type 2 diabetes patients using soft computing techniques, in: 2015 Fifth International Conference on Advances in Computing and Communications (ICACC), September, IEEE, 2015, pp. 371–375.

K. Saravananathan, T. Velmurugan, Analyzing diabetic data using classification algorithms in data mining, Indian J. Sci. Technol. 9 (43) (2016), https://doi.org/10.17485/ijst/2016/v9i43/93874.

Y. Song, J. Liang, J. Lu, X. Zhao, An efficient instance selection algorithm for k nearest neighbor regression, Neurocomputing 251 (2017) 26–34.

V.S.R.P. VarmaKamadi, RaoAllamAppab, ThummalaSitaMahalakshmia, A computational intelligence technique for the effective diagnosis of diabetic patients using principal component analysis (PCA) and modified fuzzy SLIQ decision tree approach, Appl. Soft Comput. 49 (2016) 137–145.

V.V. Vijayan, C. Anjali, Prediction and diagnosis of diabetes mellitus—a machine learning approach, in: 2015 IEEE Recent Advances in Intelligent Computational Systems (RAICS), December, IEEE, 2015, pp. 122–127.

YoichiHayashi, ShonosukeYukita, Rule extraction using recursive-rule extraction algorithm with J48graft combined with sampling selection techniques for the diagnosis of type 2 diabetes mellitus in the Pima Indian dataset, Inf. Med. Unlocked 2 (2016) 92–104.

http:/www.who.int/mediacentre/factsheets/fs312/en/.

IoT-based smart mirror for health monitoring

7

Ilango Krishnamurthy[a], **D. Prabha**[a], **M.S. Karthika**[b]

[a]Computer Science and Engineering, Sri Krishna College of Engineering and Technology, Coimbatore, India [b]Information Technology, Bannari Amman Institute of Technology, Sathyamangalam, Coimbatore, India

7.1 Introduction

Many different techniques are available for health monitoring and different types of sensors are used for such purposes. Early health monitoring techniques were clinical, wherein with the help of trained therapists an individual could learn to control specific physiological functions by changing the thoughts and perceptions that produced them, but individuals had to travel to specific locations where the technology was. Today, individuals can use biomedical sensors to keep track of physiological functions remotely, through the Internet of Things (IoT). Sensors are added to devices such as mobile phones, watches, etc. This technology is very useful for those who are able to use such devices. However, elderly people, who tend to be less able to use such devices, need health monitoring systems more than other age groups. This chapter introduces an effective health monitoring system using mirrors, a simple device used by all people regardless of age. The biomedical sensors in the mirror collect the physiological data and it is sent to medical staff to provide them with information about the patient's health condition. Using this technique, doctors can monitor their patient's health remotely.

7.2 Literature review

Yong et al. [1] report on the design of a smart mirror based on the Raspberry PI series of small single-board computers. Their mirror is designed such that the Raspberry Pi board is connected to the network through WiFI, where it obtains information such as the weather forecast, time, date, and other information, and then the information is displayed on a plasma display. The user can interact with a mobile phone through the mirror. The advantages of the intelligent mirror are its small size, low cost, simple operation, and suitability for families.

Systems Simulation and Modeling for Cloud Computing and Big Data Applications
https://doi.org/10.1016/B978-0-12-819779-0.00007-1

Hossain et al. [2] also published a paper on the design of smart mirror using Raspberry Pi. The paper gives details on the design and development of an interactive multimedia futuristic smart mirror with artificial intelligence (AI) for the home environment and also for commercial uses in different industries. Here interaction with the mirror is possible through speech-processing technology.

The work of Henriquez et al. [3] discusses a smart mirror that can detect and monitor facial signs over time, correlating them with cardio-metabolic risk and thereby providing guidance to users on how to improve their habits. The paper describes the European project SEMEOTICONS to develop a device to detect and monitor facial signs, thereby giving personal guidance for lifestyle. The clinical validation of the Wize mirror is ongoing. It focuses on the reproducibility of measurements provided by the Wize mirror and the correlation of estimated wellness with cardio-metabolic risk charts.

García et al. [4] report on a smart mirror that uses authentication with facial recognition and user information to provide a custom news report according to the user profile. The facial recognition authentications, customized news recommendation system, and voice commands are the main added features to the existing system.

Gomez-Carmona et al. [5] discuss a multiuser smart mirror system to promote wellness and healthier lifestyles in the work environment. By using an RFID reader, the mirror recognizes different users through their personal ID card and thereby allows them to access a personalized user interface. The mirror provides the environmental conditions in the workplace, personal information obtained from wearable devices, and general information also. Employees can also access their ranking position from the mirror.

Kaur et al. [6] present information showing that the IoT can play a significant role in remote health monitoring. In their paper they propose a system for monitoring the pulse rate and body temperature of a person by using dedicated sensors along with a Raspberry Pi and the IoT. Remote monitoring is achieved by storing the collected data to the Bluemix Cloud (now IBM Cloud). Then the data can be accessed by doctors from anywhere in the world. It uses a temperature sensor and heart rate sensor to collect the physiological data of the user.

Asthana et al. [7] address the problem of proactive monitoring of one's health by recommending wearable technologies and IoT solutions that can be used by individuals. This can be achieved by analyzing the individual's unstructured medical history using text mining, adding this data to structured demographic attributes and then giving this data to a machine learning classification model which predicts diseases. Then it maps these diseases to the attributes and finally it uses a mathematical optimization model to suggest wearable devices to the user.

Lakshmanachari et al. [8] propose an android-based mobile data acquisition solution that collects the personalized health information of the end user and then stores, analyzes, and visualizes it on smart devices. The smart mobile device can collect information from various wireless and wired sensors. Embedded sensors in the mobile device provide useful status information also. This solution provides low complexity, low power consumption, and a highly portable system for health monitoring.

Sundaravadivel et al. [9] report on current trends and also the challenges and opportunities available in smart healthcare. They say that, by reducing the gap between researchers and healthcare professionals, more research problems and diseases can be addressed and smarter lifestyles can be adopted.

Gómez et al. [10] discuss an architecture based on an ontology that is capable of monitoring health and recommend eating habits and workout routines to patients. People with diseases like diabetes, heart disease and high blood pressure, among others, are more recommended to use this technique.

Athira et al. [11] propose a smart mirror which used for home automation. It controls the mechanisms of home appliances and the opening and closing of cabinets through speech recognition techniques. It have some additional functionality, like a reminder service by mobile synchronization and through social media.

Gold et al. [12] describe SmartReflect, which is a smart platform for developing smart mirror applications. The main features are that it is modular, lightweight, and extensible. It allows developers to sidestep the sandboxed environment created by web browsers and it support plugins written in any programming language.

Liu et al. [13] compare smart sensors, smart objects, and the things in the IoT. Both similarities and differences have been identified. Comparisons have been made for definitions provided by various authors and organizations.

Sivakanth and Kolangiammal [14] describe a system designed to monitor the temperature and heartbeat of a patient using the IoT. Real-time information is sent to several users including doctors in critical situations and also a buzzer is used to alert the caretaker.

Kumar and Rajasekaran [15] also describe a health monitoring system using the IoT. The sensors will gather the medical information of the patient, which includes heart rate and blood pressure. It is then sent to the Internet through Raspberry Pi, so a doctor can monitor the health condition of the patient from any place in the world.

7.3 **Benchmarks**

Product	Features	Advantage	Disadvantage
Magic mirror	It comes with the basic widgets for time, calendar, weather and news	It looks very good and has a clean user interface	It has no physical interaction with the user and simply acts as an information panel
Home mirror	It also shows useful information like time, weather, date, and reminders	It uses an android tablet behind the mirror, so no need of a separate computer board and screen	It also lacks any kind of interaction

Continued

Product	Features	Advantage	Disadvantage
Evan Cohen's smart mirror	It offers a wide range of voice commands to interact with the mirror	The voice commands support time, date, and weather information, showing maps, adding reminders, etc.	Voice commands cannot be sufficient to control other devices like home automation
Max Braun's smart mirror	It updates the useful information automatically	It supports voice commands to perform Google search	It has a small level of interactivity, but still is basically an information panel only
PANL	It is one of the first touchscreen smart mirrors	It can display useful information like time, date, weather, and also can play music, YouTube videos, etc.	The founder has not revealed much information about the hardware and software part of the mirror

7.4 Proposed system

The proposed system is an IoT-based smart mirror that will monitor the health conditions of people by using different sensors, allowing data to be captured and analyzed for diagnosis and treatment by doctors (Fig. 7.1).

7.5 Components

7.5.1 One-way mirror

A one-way mirror, sometimes called a two-way mirror, is a mirror that is partially reflective and partially transparent. The dark or black side of the screen is seen as a reflection and the other side can be seen through.

7.5.2 Display

The smart mirror needs a display that is connected to the one-way mirror. TV screens are now used as display, but use of a tablet as a screen is more cost effective. An HDMI cable is used to connect the display with Raspberry Pi.

7.5.3 Raspberry Pi

The Raspberry Pi is a single-board computer developed by the Raspberry Foundation in the UK. It lacks a hard drive and it does not have any preinstalled operating system (OS). To install the OS a micro SD card is needed.

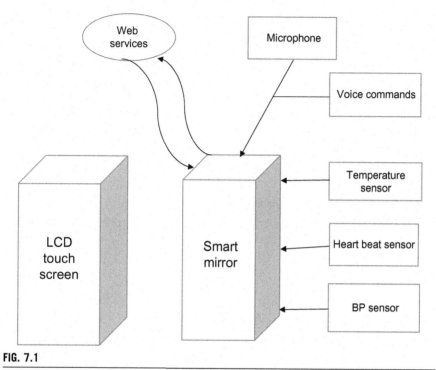

FIG. 7.1

IoT-based smart mirror for health monitoring.

7.5.4 Microphones

To interact with the mirror, microphones are used. Two microphones are used to power the voice recognition capabilities. The first microphone is a simple one connected through a USB sound card to the Pi board. The second microphone is a PS3 eye camera, which can directly connect through USB.

7.5.5 Frame and support

The mirror frame is made of wood and it provides support for the mirror and all other components.

7.5.6 Sensors

The system uses different biomedical sensors to read the human physiological information in order to monitor the health condition.

A handheld ECG sensor is used to provide ECG biofeedback information. The result is used to monitor the heart activity patterns of the user. Measuring the stress level is one of the objectives of recording the ECG signal.

The blood pressure sensor measures the systemic arterial pressure at the user's finger. It detects the blood flow and thereby is able to measure blood volume changes in the finger.

A skin temperature sensor measures the palm temperature of the user. There are different sensors available that can measure the skin temperature and humidity. Such measurements will increase the understanding of the user's health condition.

A digital scale is also used to measure the weight of the user continuously. The readings obtained are helpful to those who are trying to lose weight.

7.6 Modules

7.6.1 Mirror surface module

A one-way mirror is attached to the display on which the sensors and all other modules are contained.

7.6.2 Sensing module

Different biomedical sensors such as temperature sensor, handheld ECG, and BP sensors are added to the mirror surface to collect the user's physiological information.

7.6.3 Storage module

The data collected from different biomedical sensors will be updated in the cloud.

7.6.4 Communication module

The natural communication between the user and the mirror can be achieved through microphones and ultrasonic sensors.

7.7 Conclusion

The proposal encompasses a Raspberry Pi 2 kit that facilitates health monitoring for people using different biomedical sensors to collect physiological features such as heartbeat, blood pressure, body temperature, etc. and sends the data to the corresponding doctors for further interpretation and handling.

References

[1] S. Yong, L. Geng, K. Dan, Design of smart mirror based on Raspberry Pi, in: 2018 International Conference on Intelligent Transportation, Big Data & Smart City (ICITBS), IEEE, 2018.

[2] M.A. Hossain, P.K. Atrey, A. El Saddik, Smart mirror for ambient home environment, in: 2007 3rd IET International Conference on Intelligent Environments, IEEE, 2007, pp. 589–596.

[3] P. Henriquez, et al., Mirror mirror on the wall... an unobtrusive intelligent multisensory mirror for well-being status self-assessment and visualization, IEEE Trans. Multimedia 19 (7) (2017) 1467–1481.

[4] I.C.A. García, et al., Implementation and customization of a smart mirror through a facial recognition authentication and a personalized news recommendation algorithm, in: 2017 13th International Conference on Signal-Image Technology & Internet-Based Systems (SITIS), IEEE, 2017.

[5] O. Gomez-Carmona, D. Casado-Mansilla, SmiWork: an interactive smart mirror platform for workplace health promotion, in: 2017 2nd International Multidisciplinary Conference on Computer and Energy Science (SpliTech), IEEE, 2017.

[6] A. Kaur, A. Jasuja, Health monitoring based on IoT using Raspberry Pi, in: 2017 International Conference on Computing, Communication and Automation (ICCCA), IEEE, 2017.

[7] S. Asthana, A. Megahed, R. Strong, A recommendation system for proactive health monitoring using IoT and wearable technologies, in: 2017 IEEE International Conference on AI & Mobile Services (AIMS), IEEE, 2017.

[8] S. Lakshmanachari, et al., Design and implementation of cloud based patient health care monitoring systems using IoT, in: 2017 International Conference on Energy, Communication, Data Analytics and Soft Computing (ICECDS), IEEE, 2017.

[9] P. Sundaravadivel, et al., Everything you wanted to know about smart health care: evaluating the different technologies and components of the internet of things for better health, IEEE Consum. Electron. Mag 7 (1) (2018) 18–28.

[10] J. Gomez, B. Oviedo, E. Zhuma, Patient monitoring system based on internet of things, Procedia Comput. Sci. 83 (2016) 90–97.

[11] S. Athira, et al., Smart mirror: A novel framework for interactive display, in: 2016 International Conference on Circuit, Power and Computing Technologies (ICCPCT), IEEE, 2016.

[12] D. Gold, D. Sollinger, Smartreflect: a modular smart mirror application platform, in: 2016 IEEE 7th Annual Information Technology, Electronics and Mobile Communication Conference (IEMCON), IEEE, 2016.

[13] X. Liu, O. Baiocchi, A comparison of the definitions for smart sensors, smart objects and things in IoT, in: 2016 IEEE 7th Annual Information Technology, Electronics and Mobile Communication Conference (IEMCON), IEEE, 2016.

[14] T.S. Sivakanth, S. Kolangiammal, Design of IoT based smart health monitoring and alert system, Int. J. Circuit Theory Appl. 9 (15) (2016) 7655–7661.

[15] R. Kumar, M.P. Rajasekaran, An IoT based patient monitoring system using Raspberry Pi, in: IEEE International Conference in Computing Technologies and Intelligent Data Engineering (ICCTIDE), January, 2016, pp. 1–4.

Discovering human influenza virus using ensemble learning

M. Nandhini, M.S. Vijaya

PSGR Krishnammal College for Women, Coimbatore, India

8.1 Introduction

Swine influenza is an infection caused by one of several types of swine influenza viruses. The swine flu is also seen in humans, where it is caused by human influenza viruses. The influenza virus is an RNA virus and comprises three types: influenza A viruses, influenza B viruses, and influenza C viruses. The influenza A virus is subdivided into H1N1 virus, H3N2 virus, H7N7 virus, H1N2 virus, H9N2 virus, and H7N2 virus, based on two proteins on the surface of the virus. Mutations on these proteins may result in different influenza subtypes [1]. Influenza A viruses can infect people, birds, pigs, horses, seals, and whales.

The two proteins are hemagglutinin (H) and neuraminidase (N). Hemagglutinin has 18 subtypes, and neuraminidase has 11 subtypes. The subtypes of hemagglutinin are H1 to H18, and the subtypes of neuraminidase are N1 to N11. Consider the H7N2 viruses, which represent an influenza A subtype that has an HA7 protein and NA2 protein. The H5N1 viruses represent the HA5 protein and NA1 protein. Influenza B virus are normally found only in humans and these viruses are not divided into subtypes. Influenza C viruses cause mild illness in humans and they also are not divided into subtypes. Two different processes are performed, namely genetic drift and genetic shift. The genetic drift process often results in different strains of H1N1 and H3N2 circulating in humans during annual influenza seasons. The genetic shift process undergoes infrequent and sudden changes of genome segments from different viral strains, which is speculated to be the major cause for influenza pandemics [1].

The H1N1 is currently pandemic in both human and swine populations. The H1N2 is currently endemic in both human and swine populations. The hemagglutinin protein of the H1N2 virus is similar to that of the currently circulating H1N1 viruses, and the neuraminidase protein is similar to that of the current H3N2 viruses. The H3N2 is currently endemic in both human and swine populations. It evolved from H2N2 by antigenic shift and caused the Hong Kong flu pandemic of 1968 and 1969 that killed up to 750,000 people. The dominant strain of annual flu in January 2006 was H3N2. Measured resistance to the standard antiviral drugs amantadine and rimantadinein H3N2 increased from 1% in 1994 to 12% in 2003 to 91% in 2005.One person in New York in 2003 and one person in Virginia in 2002 were found to have serologic evidence of infection with H7N2. H7N7 has unusual zoonotic potential. In 2003 in The Netherlands, 89 people were confirmed to have H7N7 influenza virus infection following an outbreak in poultry on several farms. Low pathogenic avian influenza A (H9N2) infection was confirmed in 1999, in China and Hong Kong in two children, and in 2003 in Hong Kong in one child. Table 8.1 shows the characteristics of influenza virus.

Virus prediction is the process of identifying the virus that causes a particular disease. Early identification of a virus helps to prevent humans from being infected and aids vaccine development. Diagnosis is usually based on signs, symptoms, and physical examination of a patient. The diagnosis of a disease and identifying the presence of a virus requires vigilant and accurate analysis. The presence of a virus can be detected by analyzing the composition of its protein sequence.

To solve this problem, several computational techniques are being adopted in the existing research. Machine learning is a field of artificial intelligence concerned with the development of algorithms that allow computers to learn and this technology has found many applications. Hence it is proposed to develop a virus prediction model by formulating the human influenza virus disease identification problem as a classification task and build it using ensemble learning.

8.2 Literature survey

Currently several methodologies have been adopted in research related to works in virus recognition using biological sequencing. This background study is carried out to understand the nature of the work. The need for the proposed work is identified

Table 8.1 Characteristics of influenza virus.

Characteristics	Influenza A	Influenza B	Influenza C
Number of genomic segments	8	8	7
Number of viral proteins	10	11	9
Reassortment	Present	Absent	Absent
Species affected	Humans, swine, equine, avian, canines, bats, marine mammals	Humans	Humans, swine
Potential to cause pandemics	Yes	No	No
Present in annual seasonal influenza vaccines	Yes	Yes	No

based on the literature survey. This section explains different research studies related to the proposed work. The literature survey describes information about the research problem undertaken, algorithms or techniques used, and data collection. A review of some research work related to virus classification is given in the following paragraphs.

In 2009, Jianmin Ma, Minh N. Nguyen, and Jagath C. Rajapakse developed a model to classify the types of HLA (human leukocyte antigen) genes by codon usage bias. HLA genes were extracted from the IMGT/HLA sequence database. A total of 1841 gene sequences were used for gene classification. A model was implemented using binary and multiclass SVM. Binary SVM achieved an accuracy rate of 99.3% and multiclass SVM achieved an accuracy rate of 99.73%. The performance of the classifiers was compared in terms of mean, standard deviation, and accuracy [2].

In 2015, Marimuthu Thangam and Balamurugan Vanniappan proposed a new approach based on periodic association rules (PARs) to predict the type of dengue virus. The dengue gene sequences datasets of the National Center for Biotechnology Information (NCBI) (United States) were used. The RECFIN algorithm finds the palindrome in the dengue gene sequence, which helps to analyze the formation of proteins. The analysis was made on three types of periodicity, namely element, subsequence, and latent periodicity. The proposed work compared the RECFIN algorithm results with the NCBI-BLAST algorithm [3].

In 2017 Fayroz F. Sherif, Nourhan Zayed, and Mahmoud Fakhr developed a model to classify the types of host origin using HA and NA proteins and feature vectors as input to the model. Protein sequences were downloaded from NCBI's influenza virus resources. The features were extracted from protein sequences using two methods: amino acid composition (AAC) and physicochemical properties (composition, transition, and distribution). A model was created using random forest and K-nearest neighbor (KNN) and it achieved an accuracy of 96.6% [4].

In 2018 S. Anitha, M. Suganthi and T.S. Gnanendra proposed a model to predict the protein associated with glaucoma by amino acid composition feature as input to machine learning methods. The protein sequences were extracted from the Swiss-Prot database. The five-fold cross-validation technique was used for increasing the performance of the classification models. A model was developed using the Weka tool. The classifiers were compared in terms of sensitivity, specificity, and accuracy [5].

The authors Pavan K. Attaluri, Ximeng Zheng, Zhengxin Chen and Guoqing Lu employed a machine learning approach to identify the host origin of human influenza H1N1 viral strains. A model was built using SVM and decision tree methods through virus sequences as inputs [1].

In building an accurate classification model through machine learning, the features describing the pattern of the sequence are highly important. The existing research work focused on prediction of viruses uses amino acid composition, transition, distribution, element periodicity, subsequence periodicity, and latent periodic features of gene sequences. But only in a few cases were protein sequences used and some of the features like molecular weight (MW), pI (isoelectric point), lengthpep, and Z-scales were not considered for learning the classifier. Hence it is proposed in this paper to build a model for virus classification using ensemble learning by capturing discriminative features from protein sequences of the influenza virus. Nine types of influenza viruses have been taken into account and features such as amino acid composition, composition, transition, distribution, molecular weight, pI, length-pep, and Z-scales, and also element, subsequence and latent periodicity, are captured to learn the classifier.

8.3 Proposed work

The aim of the proposed work is to implement an efficient ensemble learning classifier for predicting the human influenza virus based on protein sequences and its contributive features. The framework of proposed methodology comprises data collection, preprocessing, feature extraction, and building the classifier using ensemble learning. Various building blocks of the proposed methodology are depicted in Fig. 8.1 and described in the following section.

8.3.1 Data collection

Proteins are made up of linear sequences of 20 amino acids joined together by peptide bonds. The 20 amino acids are alanine, arginine, asparagine, aspartic acid, cysteine, glutamic acid, glutamine, glycine, histidine, isoleucine, leucine, lysine, methionine, phenylalanine, proline, serine, threonine, tryptophan, tyrosine, and

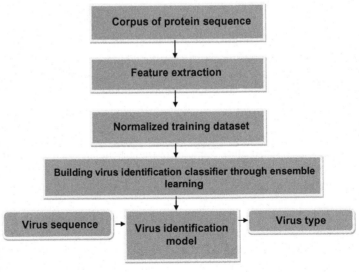

FIG. 8.1

Viruses identification model.

valine. The letter codes representing each amino acid is A, R, N, D, C, E, Q, G, H, I, L, K, M, F, P, S, T, W, Y, V.

The protein sequences associated with a virus of a human host were extracted from the NCBI's influenza virus resources. A total of 404 sequences were selected for nine virus types: influenza A virus, influenza B virus, influenza C virus, H1N1 virus, H1N2 virus, H3N2 virus, H7N2 virus, H7N7 virus, and H9N2 virus. Fifty sequences were collected from influenza A virus, influenza B virus, H7N7, and H9N2 virus. For the influenza C virus, 45 sequences were collected; for H1N1, H1N2, and H3N2, 49 sequences in each case were collected; and 12 sequences for the H7N2 virus were retrieved. These virus sequences were stored as fasta files. Table 8.2 summarizes the count of protein sequences associated with the influenza viruses. A sample of the protein sequence is given as follows.

>AOK93137.1 neuraminidase [Influenza A virus (A/Alappey/MCVR449/2009 (H3N2))] MNPNQKIITIGSVSLTISTICFFMQTAILITTVTLHFKQCEFNSPPNN QVMLCEPTIIERNITEIVYLTNTTIEKEICPKLAEYRNWSKPQCDITGFAPFSK DNSIRLSAGGDIWVTREPYVSCDPDKCYQFALGQGTTLNNVHSNNTVRDR TPYRTLLMNELGVPFHLGTKQ VCIAWSSSSCHDGKAWLHVCITGDDKNA-TASFIYNGRLVDSVVSWSKEILRTQESECVCINGTCTVVMTDGSASGKADT-KILFIEEGKIVHTSTLSGSAQHVEECSCYPRYPGVRCVCRDNWKGSNRPIV-DINIKDHSIVSSYVCSGLVGDTPRKNDSSSSSHCLDPNNEEGGRGVKGWAF DDGNDVWMGRTISEKSRLGYETFKVIEGWSNPKSKLQINRQVIVDRGNRS-GYSGIFSVEGKSCINRCFYVELIRGRKEETEVLWTSNSIVVFCGTSGTYGTG SWPDGADINLM PI.

8.3.2 Feature extraction

Feature extraction plays an important role in enhancing the performance of the classification models. In this research work, amino acid composition, physicochemical properties (composition, transition, and distribution), element periodicity, subsequence periodicity, latent periodicity, Z-scales, lengthpep, pI, and molecular weight

Table 8.2 Counts of protein sequences.

Virus name	No. of sequences
Influenza A virus	50
Influenza B virus	50
Influenza C virus	45
H1N1 virus	49
H1N2 virus	49
H3N2 virus	49
H7N2 virus	12
H7N7 virus	50
H9N2 virus	50

are considered as essential features for describing a virus sequence. A set of 20 features has been derived for element periodicity and amino acid composition. From subsequence periodicity, 9 features, and from latent periodicity, 4 features have been extracted. From composition, transition, and distribution, a group of 21 features and for Z-scales, a group of 5 features have been derived. A single feature was derived for pI, MW, and lengthpep. These features are derived from protein sequences and are given in the following paragraphs.

8.3.2.1 Element periodicity

The element periodicity is calculated by counting the repeated number of each amino acid of the protein sequence. The values for the 20 dimensions A,C,D,E,F,G,H,I,K,L, M,N,P,Q,R,S,T,V,W,Y have been computed. For the sample sequence shown earlier, the element periodicity values are given as:

$A=14$, $C=22$, $D=24$, $E=26$, $F=15$, $G=38$, $H=9$, $I=38$, $K=24$, $L=25$, $M=7$, $N=30$, $P=19$, $Q=12$, $R=23$, $S=47$, $T=37$, $V=35$, $W=11$

8.3.2.2 Subsequence periodicity

The subsequence periodicity is calculated by counting the repeated number of amino acids with specific length. In subsequence with length 2, the values for seven dimensions namely S2-1, S2-2, S2-3, S2-4, S2-5, S2-6 and S2-7 have been computed. For length 3, the values for two dimensions, S3–1 and S3–2, have been calculated. The subsequence periodicity values for the sample sequence are given as:

S2-1$=$108, S2-2$=$37, S2-3$=$15, S2-4$=$2, S2-5$=$0, S2-6$=$0, S2-7$=$0
S3-1$=$153, S3-2$=$2

8.3.2.3 Latent periodicity

Latent periodicity is used to find the presence of a palindrome in the protein sequence with specific length. For length 2, the values for two dimensions such as False-P2 and True-P2 have been computed. For length 3, the values for two dimensions, False-P3 and True-P3, have been computed. For the sample sequence shown earlier, the latent periodicity values are given as:

FALSE-P2$=$215, TRUE-P2$=$20
FALSE-P3$=$148, TRUE-P3$=$9

8.3.2.4 Amino acid composition

Amino acid composition is calculated by the amino acid function, which produces 20 amino acid values. In amino acid composition, the values for 20 dimensions representing the amino acids A, R, N, D, C, E, Q, G, H, I, L, K, M, F, P, S, T, W, Y and V have been calculated. The amino acid composition values for the sample sequence are given as:

A=0.029850746	R=0.049040512	N=0.063965885	D=0.051172708
C=0.046908316	E=0.0554371	Q=0.025586354	G=0.081023454
H=0.019189765	I=0.081023454	L=0.053304904	K=0.051172708
M=0.014925373	F=0.031982942	P=0.040511727	S=0.10021322
T=0.078891258	W=0.023454158	Y=0.02771855	V=0.074626866

8.3.2.5 Composition/transition/distribution

In amino acid physicochemical properties, three groups and seven attributes are used. The group consists of composition, transition, and distribution computed by the CTD function, which gives seven attribute values: hydrophobicity, polarizibility, normalized van der Waals, secondary structure, charge, solvent accessibility, and polarity. In each case of hydrophobicity, normalized van der Waals, polarity, polarizability, secondary structure, solvent accessibility and charge, the values for 21 dimensions, namely HG1, HG2, HG3, NG1, NG2, NG3, PG1, PG2, PG3, PZG1, PZG2, PZG3, SSG1, SSG2, SSG3, SAG1, SAG2, SAG3, CG1, CG2, and CG3, have been calculated. For the sample sequence shown earlier, the composition, transition, and distribution values are given as:

HG1=0.296375267	HG2=0.377398721
HG3=0.326226013	NG1=0.428571429
NG2=0.353944563	NG3=0.217484009
PG1=0.353944563	PG2=0.330490405
PG3=0.315565032	PZG1=0.341151386
PZG2=0.441364606	PZG3=0.217484009
CG1=0.10021322	CG2=0.793176972
CG3=0.106609808	SSG1=0.298507463
SSG2=0.364605544	SSG3=0.336886994
SAG1=0.42217484	SAG2=0.296375267
SAG3=0.281449893	

8.3.2.6 Lengthpep

The lengthpep is a total count of amino acid in a sequence and the value for the single dimension, called LP, is computed by the lengthpep function. The length peptide value for the sample sequence is given as:

LP=469

8.3.2.7 Molecular weight

The molecular weight is a weight of amino acids in sequence and the value for the single dimension, called MW, is calculated by the mw function. The molecular weight value for the sample sequence is given as:

MW=51,972.39

8.3.2.8 pI

p*I* refers to the isoelectric point and the value for the single dimension, called p*I*, is calculated by the p*I* function. For the sample sequence shown earlier, the p*I* value is given as:

p*I* = 6.737443

8.3.2.9 Z-scales

Z-scales are based on physicochemical properties of amino acids such as lipophilicity, steric properties, electronic properties, electronegativity, and heat of formation, etc., and the values for 5 dimensions, namely Z1, Z2, Z3, Z4, and Z5, have been computed by the z-scales function. The Z-scales values for the sample sequence are given as:

Z1 = 0.310469, Z2 = −0.54851, Z3 = −0.11354, Z4 = −0.48373 and Z5 = 0.105821

8.3.3 Dataset

The features extracted from protein sequences are transformed into feature vectors to train the ensemble learning algorithms. The corpus consists of 404 protein sequences of 9 virus types. The dataset with 404 instances of dimension 84 has been created and the class labels are assigned as A, B, C, H1N1, H1N2, H3N2, H7N2, H7N7, and H9N2 for the corresponding nine virus types. Finally, the dataset is normalized using minmax normalization to improve the performance of the classification models.

8.3.4 Ensemble learning classifiers

The ensemble learning methods combine the predictions of multiple learning algorithms. This provides enhanced predictive performance over a single learner. In this research work the viruses have been classified using ensemble learning techniques. The classification models have been built using bagging, boosting, voting, and randomized tree classifiers. In the bagging classifier, a bagging algorithm is used. The boosting classifier consists of two types of algorithms, AdaBoost and gradient boosting. In this work, the AdaBoost algorithm is implemented. The randomized tree classifier comprises two types, random forest and extremely randomized trees. Here, the extremely randomized trees algorithm is applied in another experiment. The voting classifier includes two types: soft voting and hard voting. SVM gridsearch soft voting is used in the fourth experiment.

8.4 Experiments and results

The human influenza virus prediction model is developed using scikit-learn, a python library for machine learning. Nine types of influenza virus were considered for implementing the model and thus the virus recognition problem becomes multiclassification. The preceding dataset was divided into training and test datasets, with 80% of the instances constituting training and 20% of the data for testing. The ensemble learning methods, specifically bagging, boosting, randomized trees and voting, were used with a training dataset to train the models through scikit-learn. The ensemble classifiers were evaluated for their efficiency using the test dataset. Metrics such as precision, recall, F1-score and accuracy were used to assess the classifier's output quality.

The virus classification model based on the bagging classifier produced 73% accuracy. The performance results of the bagging classifier in virus prediction in terms of various metrics are shown in Tables 8.3 and 8.4.

The virus classification model based on the boosting classifier produces 75% predictive accuracy and the performance results of the boosting classifier in terms of various metrics are shown in Tables 8.5 and 8.6.

The virus classification model based on the randomized trees classifier yields 75% accuracy. The performance results of the randomized trees classifier in virus prediction in terms of various metrics are shown in Tables 8.7 and 8.8.

Table 8.3 Performance measures of bagging classifier for nine virus types.

Measures	Values
Precision	0.6532
Recall	0.6654
F1-score	0.6560
Accuracy	73%

Table 8.4 Performance measures of bagging classifier for each virus type.

Virus name	Precision	Recall	F1-score
H1N1 virus	0.38	0.50	0.43
H1N2 virus	0.83	1.00	0.91
H3N2 virus	1.00	1.00	1.00
H7N2 virus	0.00	0.00	0.00
H7N7 virus	0.67	0.60	0.63
H9N2 virus	0.80	0.80	0.80
Influenza A virus	0.29	0.20	0.24
Influenza B virus	0.91	1.00	0.95
Influenza C virus	1.00	0.89	0.94
Avg/total	0.71	0.73	0.72

Table 8.5 Performance measures of adaboost algorithm for nine virus types.

Measures	Values
Precision	0.6996
Recall	0.6876
F1-score	0.6919
Accuracy	75%

Table 8.6 Performance measures of adaboost classifier for each virus type.

Virus name	Precision	Recall	F1-score
H1N1 virus	0.42	0.50	0.45
H1N2 virus	1.00	1.00	1.00
H3N2 virus	1.00	1.00	1.00
H7N2 virus	0.00	0.00	0.00
H7N7 virus	0.78	0.70	0.74
H9N2 virus	0.73	0.80	0.76
Influenza A virus	0.38	0.30	0.33
Influenza B virus	1.00	1.00	1.00
Influenza C virus	1.00	0.89	0.94
Avg/total	0.77	0.75	0.76

Table 8.7 Performance measures of extremely randomized trees algorithm for nine virus types.

Measures	Values
Precision	0.7078
Recall	0.6888
F1-score	0.6917
Accuracy	75%

Table 8.8 Performance measures of extremely randomized trees for each virus type.

Virus name	Precision	Recall	F1-score
H1N1 virus	0.43	0.60	0.50
H1N2 virus	1.00	1.00	1.00
H3N2 virus	0.91	1.00	0.95
H7N2 virus	0.00	0.00	0.00
H7N7 virus	0.70	0.70	0.70
H9N2 virus	1.00	0.70	0.82
Influenza A virus	0.33	0.20	0.25
Influenza B virus	1.00	1.00	1.00
Influenza C virus	1.00	1.00	1.00
Avg/total	0.77	0.75	0.76

The virus classification model based on voting classifier gives 74% accuracy and the performance results of the voting classifier in terms of various metrics are shown in Tables 8.9 and 8.10.

Table 8.9 Performance measures of SVM gridsearch soft voting algorithm for nine virus types.

Measures	Values
Precision	0.6616
Recall	0.6777
F1-score	0.6665
Accuracy	74%

Table 8.10 Performance measures of SVM gridsearch soft voting algorithm for each virus type.

Virus name	Precision	Recall	F1-score
H1N1 virus	0.42	0.50	0.45
H1N2 virus	0.83	1.00	0.91
H3N2 virus	1.00	1.00	1.00
H7N2 virus	0.00	0.00	0.00
H7N7 virus	0.55	0.60	0.57
H9N2 virus	0.88	0.70	0.78
Influenza A virus	0.38	0.30	0.33
Influenza B virus	0.91	1.00	0.95
Influenza C virus	1.00	1.00	1.00
Avg/total	0.72	0.74	0.73

8.4.1 Comparative analysis

The performances of all the four ensemble classifiers are compared against their evaluation metrics. From comparative analysis, it is observed that the AdaBoost and extremely randomized trees classifiers gives more effective results than bagging and SVM gridsearch soft voting for recognizing virus from the protein sequence. From analyzing the results, it is perceived that the AdaBoost and extremely randomized trees prediction model show a higher accuracy of 75% than the bagging and SVM gridsearch soft voting.

The other two classifiers, namely bagging and SVM gridsearch soft voting, give an accuracy of 73% and 74%. The precision and recall are higher for extremely randomized trees than the other algorithms. The F1-score is higher in extremely randomized trees and AdaBoost when compared to the other learning methods. Table 8.11 shows the comparative results of ensemble learning classifiers in terms of accuracy, precision, recall, and F1-score, as illustrated in Fig. 8.2.

Table 8.11 Comparative results of the classifiers.

Evaluation criteria	Classifiers			
	Bagging	**AdaBoost**	**Extremely randomized trees**	**SVM gridsearch soft voting**
Precision	0.6532	0.6996	0.7078	0.6616
Recall	0.6654	0.6876	0.6888	0.6777
F1-score	0.6560	0.6919	0.6917	0.6665
Accuracy	73%	75%	75%	74%

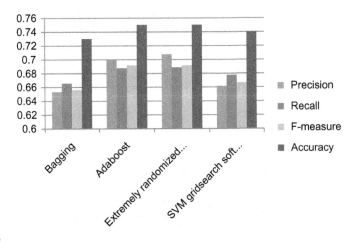

FIG. 8.2

Comparison of all ensemble classifiers.

8.5 Conclusion

Accurate detection of the human influenza virus can significantly improve influenza surveillance and vaccine development. This chapter demonstrates discovering the human influenza virus using protein sequences through ensemble learning as a multiclassification task. The protein sequences were collected from the NCBI Influenza Virus Sequence Database and the appropriate dataset was developed for supervised learning. The virus prediction models were generated using ensemble classification and the results were analyzed. The analysis of experimental results shows that the boosting and randomized trees-based virus recognition models are suited for predicting the type of virus. The research work can be further extended by enhancing the dataset with more instances and dimensions and experimenting with advanced learning techniques like deep learning.

References

[1] P.K. Attaluri, X. Zheng, Z. Chen, G. Lu, Applying machine learning techniques to classify H1N1 viral strains occurring in 2009 flu pandemic, 2009.

[2] J. Ma, M.N. Nguyen, J.C. Rajapakse, Gene classification using codon usage and support vector machines, IEEE/ACM Trans. Comput. Biol. Bioinform. 6 (1) (2009).

[3] M. Thangam, B. Vanniappan, Mining association rules in dengue gene sequence with latent periodicity, Comput. Biol. J. 2015 (2015) 839692. https://doi.org/10.1155/2015/839692.

[4] F.F. Sherif, N. Zayed, M. Fakhr, Classification of host origin in influenza a virus by transferring protein sequences into numerical feature vectors, Int. J. Biol. Biomed. Eng. 11 (2017).

[5] S. Anitha, M. Suganthi, T.S. Gnanendra, Evaluation of data mining classifiers for prediction and classification of glaucoma associated proteins, Int. J. Pharm. Bio. Sci 9 (1) (2018) 1–11.

Further reading

O. Almadani, R. Alshammari, Prediction of stroke using data mining classification techniques, Int. J. Adv. Comput. Sci. Appl. 9 (1) (2018).

S.M. Gorade, A. Deo, P. Purohit, Early identification of diseases based on responsible attribute using data mining, Int. Res. J. Eng. Technol. (IRJET) 4 (7) (2017).

J. Han, M. Kamber, Data Mining: Concepts and Techniques, second ed., The Morgan Kaufmann Publishers, 2006.

W.J.H. McBride, H. Bielefeldt-Ohmann, Dengue viral infections; pathogenesis and epidemiology, Microbes Infect. 2 (9) (2000) 1041–1050.

S. Neelamegam, E. Ramaraj, Classification algorithm in data mining: an overview, Int. J. P2P Netw. Trends Technol. (IJPTT) 4 (8) (2013).

Mining and monitoring human activity patterns in smart environment-based healthcare systems

M. Janani[a], M. Nataraj[b], C.R. Shyam Ganesh[c]
[a]Department of CSE, Jai Shriram Engineering College, Tirupur, India [b]Department of CSE, Kongu Engineering College, Erode, India [c]Department of CSE, PSG College of Technology, Coimbatore, India

9.1 Introduction

Data mining is the computing process of discovering patterns in large datasets involving machine learning, statistics, and database systems. It is a process of extracting data patterns by applying intelligent methods, transforming the data into an understandable structure for further use. Data mining parameters include sequence or path analysis, classification, clustering, and forecasting. The concept of data mining includes statistics, artificial intelligence, and machine learning.

A smart environment can include a smart home, smart city, smart meter, smart parking, and smart governance, as well as many sensors. Smart environments provide a way to respond to the needs of the residents in a context-aware manner. A smart environment is equipped with different types of sensors that allow the system to collect data on inhabitant activities and environmental situations. Activity in a smart home can be collected using ambient sensors such as infrared motion sensors to track the movements of residents. Additionally, ambient sensors can be temperature sensors, pressure sensors, contact switch sensors, and smart power meters, which provide context information.

A smart city is an urbanized area in which multiple sectors cooperate to achieve sustainable outcomes. This is done through the analysis of contextual and real-time information. Moreover, a smart city can provide intelligent responses to different kinds of daily needs, including livelihoods of people, security systems, public transportation and environment, public health, and industrial and commercial activities. Smart city data present multiple challenges due to their volume, velocity, and variety. The data are initially unstructured because they can be in the form of audio, images, log files, tweets, text, etc., and they must be integrated with structured data.

Systems Simulation and Modeling for Cloud Computing and Big Data Applications
https://doi.org/10.1016/B978-0-12-819779-0.00009-5

The sources of data can be legacy sources, new technology sources, Internet of Things (IoT) devices, or manual entry from humans. The IoT sensors may not be working properly at some locations, giving faulty or missing data. Thus, major preprocessing needs to be done on such data. Smart city data becomes big data, which is a technology term that describes large volumes of data, both structured and unstructured.

Big data is a collection of datasets of large or complex data, which traditional data processing methods cannot easily deal with. The challenges of big data are capturing data, data storage, data analysis, search, sharing, transfer, visualization, querying, updating, and information privacy. Other major challenges include information sharing, privacy, security, new data formats, and quality of data. Therefore, smart cities and smart homes face all the challenges that big data faces. In addition, they face challenges related to security and privacy as well. For example, smart meters can be installed to monitor activities in homes. Various appliances are monitored and are used to measure the individual appliance usage. Smart meters along with smart power plugs collect data to track people's activity on a regular basis, by classifying them regarding appliance usage.

Big data applications in healthcare organizations provide significant benefits, which include detecting diseases at an earlier stage and prescribing more easily and effectively. Advancements in tools and technology of smart cities have improved healthcare services, and the infrastructure and technology of smart cities have led to thinking beyond the limits of existing healthcare systems. The field of telemedicine has also contributed to a new and ubiquitous concept called smart health, which integrates ideas from ubiquitous computing and ambient intelligence applied to predictive, personalized, preventive, and participatory healthcare systems. Smart health is strongly connected to the concepts of wellness and wellbeing, which include large volumes of data collected from biomedical sensors. Therefore big data has genomic driven, payer provider, and social media data actuators to observe and predict a patient's physical and mental conditions.

Smart health is a nascent but promising field of study at the intersection of medical informatics, public health, and also business, alluding to intelligent healthcare services or enhanced cognitive capabilities provided through the IoT. Healthcare services are affected by the migration of people to cities, where digital transformation is a recent trend. To provide healthier environments, homes are being entirely organized using smart devices, which contribute to making the city smarter. As mentioned earlier, smart cities can generate reams of data, which needs to be processed, refined, and turned into useful structured data.

9.2 Literature review

Aftab and Chau proposed smart power plugs, a notable cyberphysical system for tracking and controlling appliance behavior. In existing smart plugs, advanced automated features such as online learning and classification and diagnosis of appliance behavior are not included, due to storage and computation limitations. Hence, the stand-alone smart plugs are developed to provide efficient classification and tracking. The IoT framework is used to track the performance and sensing of evenly distributed tasks using local memory [1].

Hao et al. proposed a paper in which activity is predicted using hardware and software with the support of smart homes. Smart homes are equipped with intelligent services for providing efficient living environments for users. Human behavior is monitored using ambient sensors, and multimodel interactions occurring the smart home are mined [2]. Feedback can then be generated for elderly people based on their recognized behavior.

A smart city is an urban innovation aimed to improve the quality of life [3]. It employs technology to bring about development on social and economic fronts. Some of the applications for smart meters are energy management, green buildings, solar usage, smart parking, intelligent traffic management, waste management, crime monitoring and management, water quality monitoring and leakage identification, and creating walkable localities and making areas less vulnerable to crime and disasters, for smart governance. Data captured in the smart city initiative is used to improve the livability of the city. These data include government data as well as social data used to analyze the sentiments of the citizens during important city events [3].

Jindal et al. proposed a paper focusing on healthcare advancements in information and communication technology that increase the number of users availing themselves of remote healthcare applications. The data collected on the patients in these applications vary with respect to volume, velocity, variety, veracity, and value. To process such a large collection of heterogeneous data is a major challenge, requiring a specialized approach. To address this issue, a new fuzzy rule-based classifier is used for handling big data along with cloud-based infrastructure called healthcare-as-a-service [4]. Another work, by Yassine et al., deals with healthcare services, which is a challenging aspect that is greatly affected by the vast influx of people to city centers. People in the cities are moving towards healthier environments, which will be a part of digital transformation in smart city services [5]. The massive volume of data generated from smart homes will provide further enhancements to smart cities.

A work by Venkatesh et al. studies smart health based on the IoT. The IoT envisions the creation of smart connected cities composed of ubiquitous environmental and user sensing with distributed, low-capacity computing that provide ample information regarding the citizens in various smart environments. People-centric information can be leveraged by providing a smart city infrastructure to improve smart health applications [6].

The work by Pramanik et al. deals with recent technology developments in big data leading to the growth of information and communication technology (ICT). They discuss how technology development brings in mobile computing, with communication playing the major role, and even ubiquitous computing provides opportunities for many government and private sectors, especially in healthcare industries.

Big data seems now to be a backbone of the healthcare industry. A three-dimensional structure of a paradigm shift is used and three broad technical branches (3T) are extracted that contribute to the promotion of healthcare systems [7].

The work carried out by Chen et al. looks at big data growth in the biomedical and healthcare areas, providing analysis of healthcare data for disease detection in elderly people. When the prediction data have errors, the healthcare details become incomplete and seemingly useless. Since each disease has unique symptoms and measures to be carried out, even healthcare data can be an outlier when there is no active prediction. Machine-learning algorithms are used for effective prediction and mining [8].

Emerging cyber-physical systems (CPSs) are examined by Hossain and Rahman, for providing services for elderly people in a smart home. The CPS user interactions are collected and sensed from within the smart home and these interactions can be used to monitor energy efficiency, among other things. The proposed work is an energy-efficient CPS for monitoring elderly people with use of technologies like cloud computing and big data [9]. Smart grid-based data collection is proposed by Liu et al. Power usage data is collected through smart meters and the data are differentiated by cost and load. Smart meter readings may vary between the on and off status [10].

Faustine et al. carried out a study of the urbanization in developing countries, considering big buildings and high power consumption areas, to measure the energy efficiency. Continuous monitoring of appliance power consumption in real-time is not an easy task. An aggregate power monitoring system is needed in smart homes [11].

Classification in smart cities is described by Moreno and Terroso. Smart cities are an application of big data that act as a major research opportunity. The contributions are the design and instantiation of an IoT-based architecture for applications of smart cities, management of energy in smart buildings, and extension of data analysis for detection of urban patterns, which can be used to improve public transport applied to the public tram service [12].

Mining and usage patterns are proposed by Schweizer and Zehnder. Usage pattern-based preferences are learned, to allow a smart home to achieve energy savings, using a frequent sequential pattern mining algorithm for real-time smart home data. This work elaborates on the comparison done on performance using various algorithms [13]. Singh et al. discuss the key elements of power consumption related to user activity in real time. Interdependencies of appliances in smart homes are still a major drawback in determining single appliance consumption. Information on single appliances is extracted using big data prediction [14].

The work by Chen et al. focuses on the ability of residents to collect household appliance usage data easily, due to the advent of widespread sensor technology. Prior studies on usage pattern discovery are mainly focused on mining patterns while ignoring the incremental maintenance of the mined results. The novel method used in this study is the dynamic correlation miner, which was developed to incrementally capture and maintain the usage correlations among appliances in a smart home environment. Furthermore, several optimization techniques are used effectively to reduce the search space [15]. Another work, carried out by Wilson et al., looks at activities, a descriptive term for the common ways households spend their time on cooking, washing, and other tasks. To generate activity time, smart meter data is used that are meaningful to households. Feedback from activities is obtained and reported for easy understanding and feature updating needed for elderly people [16].

Table 9.1 shows the parameter comparison of clustering algorithms from the literature survey.

9.3 Impulse-based Markov model

The proposed architecture for healthcare applications using smart environment big data and human activity patterns (impulse-based Markov model) is shown in Fig. 9.1. The smart meter data was obtained from a database. Then the data were preprocessed to obtain cleaned integrated data transformed for reduction. Data classification was performed once the data preprocessing was complete. The classification algorithm used was the semi-Markov model-based logistic regression. Classified data was clustered using impulse model-based clustering.

The clustered data can then be used in activity prediction and pattern analysis. The analyzed data is stored in the database. An alert is immediately generated when activity is predicted in the smart environment. Using activity prediction and frequent pattern analysis, healthcare services are provided by the system. The smart environment uses ambient and mobile sensing and the semi-Markov model along with

Table 9.1 Parameter comparison of clustering algorithms.

S. No	Clustering algorithms	Efficiency	Quality	Robust	Privacy	Security
1.	K-means [1]	High	Medium	Low	Very high	Medium
2.	Expectation maximization [3]	High	High	Nil	Nil	Medium
3.	Impulse-based model [5]	High	High	Medium	High	High
4.	Heuristic clustering [10, 12]	Low	Low	Medium	Medium	Nil
5.	Genetic algorithm [12, 13]	Medium	Medium	Low	Low	Nil

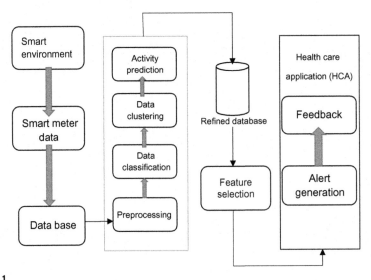

FIG. 9.1

Impulse-based Markov model.

logistic regression is used for detecting individual habitant data. Absolute predictions are determined for accuracy and performance.

A smart meter is used to measure appliance usage along with the duration. Using smart meter data, activity recognition and prediction is handled in a Bayesian network. The data is extracted and used to identify multiple appliances and energy over time is analyzed at the appliance level. Clustering of the data is used for detecting sudden changes in human activity.

Impulse model-based clustering is also used in detecting usage patterns from appliances' ON/OFF status. Clustering analysis is used to discover appliance usage based on time, also including timestamps. As mentioned, activity prediction and recognition is done in the Bayesian network, which uses a directed acyclic graph (DAG) and includes the concept of causality. The Bayesian network determines mitigating missing data, used to learn the relationship between random variables and historical data and variations while overfitting. Accuracy prediction and performance analysis are plotted as a graph in Figs. 9.2 and 9.3. User based items gain high performance and accuracy in mining and data prediction.

Data mining is used for various popular items that are commonly used appliances in smart homes. User-based items include television, washing machine, and fan, which are the most commonly used items by elderly residents. Feedback is obtained by frequent pattern mining. An alert can be generated each 30 min by appliance data fetched from the database. Feedback should be provided on the basis of active monitoring. These data can be used to improve healthcare applications. Finally, the data is integrated to build an ontology model displaying the properties and relationships among appliances.

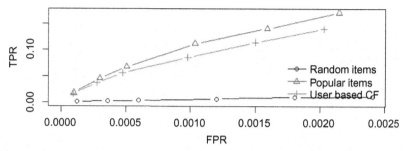

FIG. 9.2

Performance analysis.

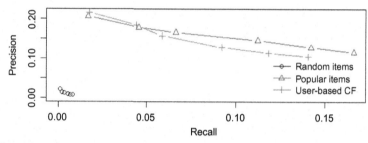

FIG. 9.3

Accuracy prediction.

9.4 **Summary and conclusion**

From the preceding comparison table, it is clear that the impulse-based model is the best-suited data clustering algorithm and the semi-Markov model based on logistic regression is best suited for data classification. The activity prediction is handled by recognizing human behavior and frequent pattern mining. Mining includes machine-learning algorithms, which may be supervised, semisupervised, and unsupervised algorithms. Activity prediction-based pattern mining is temporally analyzed using a semi-Markov model and impulse model.

The proposed model is an impulse-based Markov model in which user behavior is predicted as a service based on a single appliance applied to a real-time scenario. Multiple usage predictions are not identified by the smart meter at a single time, so smart power plugs can be used. Healthcare-related issues are identified but there is a lack of sending alerts to the patients or care providers, which is overcome in the proposed model.

An absolute model is presented for recognizing human activity patterns using smart meters, whereas pattern mining and prediction are done by user-based collaborative filtering methods. An ontology model is built to map the appliance to activity operating on any unit of time. Performance and accuracy values are obtained for random, popular, and user-based items. Items may vary with appliances. Further enhancements can be made on accuracy using other items rather than those related mostly to elderly people. The activities of elderly people are monitored in a smart home and, in case of emergency, prediction values are analyzed with mining and the results are generated to the elderly user.

References

[1] M. Aftab, C.-K. Chau, Smart power plugs for efficient online classification and tracking of application behavior, in: Proceedings of APSys, September, 2017.

[2] J. Hao, A. Bouzouane, S. Gaboury, Complex behavioral pattern mining in non-intrusive sensor-based smart environments using an intelligent activity inference engine, J. Reliab. Intell. Environ. 3 (2) (2017) 99–116.

[3] P. Joglekar, V. Kulkarni, Data oriented view of a smart city a big data approach, in: International Conference on Emerging Trends & Innovation in ICT (ICEI), July, 2017.

[4] A. Jindal, A. Dua, N. Kumar, An efficient fuzzy rule-based big data analytics scheme for providing healthcare-as-a-service, in: IEEE ICC SAC Symposium E-Health Track, July, 2017.

[5] A. Yassine, A. Alamri, Mining human activity patterns from smart home big data for health care applications, IEEE Access 5 (2017) 131–141.

[6] J. Venkatesh, B. Aksanli, C.S. Chan, Modular and personalized smart health application design in a smart city environment, IEEE Internet Things J. 5 (2017) 614–623.

[7] M.I. Pramanik, M.A.K. Azad, Smart health: big data enabled health paradigm within smart cities, Expert. Syst. Appl. 87 (2017) 370–383.

[8] M. Chen, L. Wang, Disease prediction by machine learning over big data from healthcare communities, IEEE Access 5 (2017) 869–879.

[9] M.S. Hossain, M.A. Rahman, G. Muhammad, Cyber physical cloud-oriented multi-sensory smart home framework for elderly people: an energy efficiency perspective, J. Parallel Distrib. Comput. 103 (2017) 11–21.

[10] W.M. Liu, L. Wang, Privacy preserving smart meter streaming against information leakage of appliance status, IEEE Trans. Inf. Forensics Secur. (2017).

[11] A. Faustine, S. Kaijage, K. Michael, A Survey on Non-Intrusive Load Monitoring Methodies and Techniques for Energy Disaggregation Problem, 2017. arXiv:1703.00785 [cs.OH].

[12] M. Victoria Moreno, F. Terroso-Sáenz, A. González-Vidal, Applicability of big data techniques to smart cities deployments, IEEE Trans. Ind. Inform. 13 (2) (2016) 800–809.

[13] D. Schweizer, M. Zehnder, H.F. Witschel, Using consumer behavior data to reduce energy consumption in smart homes, in: International Conference on Machine Learning and Applications, March, 2016.

[14] S. Singh, A. Yassine, S. Shirmohammadi, Incremental mining of frequent power consumption patterns from smart meters big data, in: IEEE Electrical Power and Energy Conference (EPEC), December, 2015.

[15] Y.-C. Chen, H.-C. Hung, B.-Y. Chiang, Incrementally mining usage correlations among appliances in smart homes, in: International Conference on Network Information Systems, December, 2015.

[16] C. Wilson, L. Stankovic, M. Coleman, 'Identifying the time profile of everyday activities in the home using smart meter data, in: Conference Proceedings, July, 2015.

Early detection of cognitive impairment of elders using wearable sensors

10

S. Meenakshi Ammal, L.S. Jayashree
Department of CSE, PSG College of Technology, Coimbatore, India

10.1 Introduction

Emerging trends [1] in pervasive computing, sensor technology, microelectromechanical systems (MEMS), and wireless communication techniques enable tracking of the daily activities of elders from remote locations. Advanced wireless communication technology allows the clinician to access real-time health status of patients remotely at any time. Inertial sensors [2] such as the tri-axis accelerometer, gyroscope, and magnetometer embedded in wearables can track the patient's daily activities. The inertial sensors can detect a patient's ambulatory activities and body postures. The inertial sensors and their measurements are shown in Table 10.1. The accelerometer measures the movements during human activities involving body acceleration in the x-axis, y-axis, and z-axis directions. It can detect daily activities such as standing, walking, sitting, jumping, dancing, lying, cycling, etc. The gyroscope is used to measure orientation and angular velocity. This group of inertial sensors can thus detect anomalous activity, like fall. Most of the current smartphones are embedded with accelerometers, gyroscope, magnetometer, and proximity sensors, which have the ability to detect human ambulatory activities and body postures.

The evolution of the Internet of Things (IoT) [1], edge computing, and cloud computing has facilitated a more comfortable, secure, and higher quality life for independent living elders, using a remote activity monitoring (RAM) system. Wearable sensors used in RAM systems support many services, such as detection of abnormal behavior, diagnosis of disease in early stage, and prevention of health emergencies. RAM systems can decrease healthcare costs and reduce caregiver's burdens.

Mild cognitive impairment (MCI) [3] refers to minor cognitive decline, which is a stage between aging-related memory loss and more serious cognitive decline, such as Alzheimer's disease (AD). Elders with MCI are at greater risk of developing AD than those without MCI. However, not all MCI stages progress to AD; some cases remain in the MCI stage during their lifetime. Globally, many elders aged 65 years and above suffer from MCI. As AD is an irreversible brain disorder [4], continuous monitoring

Systems Simulation and Modeling for Cloud Computing and Big Data Applications
https://doi.org/10.1016/B978-0-12-819779-0.00010-1

Table 10.1 Inertial sensors.

Inertial sensors	Measurements
Accelerometer	Acceleration
Magnetometer	Orientation
Gyroscope	Orientation

of the activities of daily living (ADLs) of MCI patients could assist in evaluating the cognitive ability of the patients and could help to prevent progression to AD. Development of cognitive impairment lowers the ability to perform ADLs such as bathing, grooming, bathroom usage, eating, going to bed, etc. The cognitive decline observed in the ADLs of the MCI patients is an early indication of progression to AD.

Actually, there is no dataset based on a dementia patient's ADLs using real-time scenario. This chapter uses the University of California, Irvine (UCI) HMP [5] dataset and artificially generates deviations in daily activities that are common symptoms found in AD patients. The dataset contains labeled records of daily activities generated using a wearable tri-axis accelerometer.

The rest of this chapter is organized as follows: Section 10.2 describes related work on human activity recognition and abnormal activity detection using sensors. Section 10.3 describes the proposed method for early detection of cognitive impairment. Section 10.4 discusses the results obtained, and Section 10.5 concludes the chapter.

10.2 Related work

Recently, human activity recognition has been employed in many applications, including medical, military, industrial, and others. A foot monitoring system [6] was designed to detect AD at an early stage using gait patterns. An IoT-based wearable device was used to collect the foot movements of individuals, and a dynamic time warping (DTW) algorithm was implemented to differentiate the foot shape of a normal person from that of an AD patient. Various machine-learning techniques such as inertial navigation algorithms, support vector machines (SVMs), and the K-nearest neighbor classifier were applied to classify the gait patterns of normal persons and AD patients. DTW generated better classification results than other machine-learning methods. The bioPLUX [7] sensor system was used to detect the symptoms of psychomotor agitation, a behavior frequently shown by AD patients. The signals from the wearable device were preprocessed using a band-pass filter to remove unwanted noise and then the extracted features were evaluated using classification algorithms. SVM gave better results in detecting psychomotor agitation. A smart-home–based abnormal activity detection system [8] was designed for patients with dementia. The activity of the patient was observed using sensors fixed within a smart home. The sensor data was processed by an activity recognition system based on a Markov logic network (MLN), to detect the patient's daily activities. The identified abnormal activity was recorded and sent to the decision support system to alert the corresponding caregivers. Based on the abnormal behavior, the decision support system prioritized the alert in terms of low alarm, high alarm, or emergency alarm.

In another system, a movement sensor and door entry point sensors [9] were placed in the home to monitor abnormal behaviors of dementia patients. The abnormal behavior was detected using the activation of the sensors from the OFF stage to ON stage and the duration of sensor activation. A recurrent neural network was applied to predict abnormal behavior in the near future.

Gait- and balance-analyzing algorithms [10] have been developed to extract the important features that are the significant indicators of AD. A wearable inertial sensor unit generates data according to the gait pattern, and the sensor data is then processed to extract the gait and balance parameters. The stride length and sway speed of healthy people and AD patients were calculated. The wearable unit automatically predicted an early stage of AD using the gait- and balance-analyzing algorithms.

In another study, gait patterns and human activities of living [11] were detected using smartphone sensors. The data collected was then passed to a computer system for further processing. The sensor data was analyzed using statistical methods and also machine-learning algorithms were applied to classify three different walking speeds: slow, normal, and fast. A cross dynamic warping (CDW) metric was applied and provided better results than other learning algorithms.

A fine-grained activity recognition system [12] was designed using multimodal wearable sensors, fixed in different body positions and generating sensor data according to body movement. The sensor data was transmitted to the server using Bluetooth beacons. A two-level supervised classifier was proposed to analyze and process the sensor data. In the first level, the sensor data from different body positions were analyzed using a modified conditional random field algorithm. In the second level, the sensor data from each wearable were fused to detect the final activity; 19 complex activities were recognized using multimodal wearable sensors.

Human activity detection and gait analysis [13] were also performed using an inertial measurement unit (IMU), fixed on the chest of the human body. The IMU has inertial sensors, including a tri-axis accelerometer, magnetometer, and gyroscope, which generate sensor data according to the daily activities of living. A continuous hidden Markov model (HMM) and circular HMM were applied to detect gait pattern and human activity such as walking, running, going upstairs, going downstairs, and standing. As the system was designed without any wireless connectivity, it restricts human mobility and the remote access of health information.

An electronic insole [14] was designed to monitor ADLs and sports activities. The electronic insole was fabricated using a temperature sensor, humidity sensor, tri-axis accelerometer, and pressure sensor. It also includes ZigBee transmission and local data storage, which permits the continuous monitoring of patient mobility and remote access of patient activities. The foot acceleration and foot orientation are used to measure ambulatory activities such as walking, walking upstairs, walking downstairs, and running. The electronic insole allows the user to walk comfortably and in a natural way. However, the different sizes of shoes on the market require the designing of different insole sizes. The LabVIEW software-based application was designed to access the sensor data via ZigBee.

A monitoring system for hand and wrist movements [15] was designed to support stroke rehabilitation. A magnetic ring and two tri-axial magnetometers were designed to measure the angular distance of wrist and finger joints. The magnetic ring was fixed on the index finger and the magnetometer was fixed on the wrist. A radial basis function network was applied to the sensor data to measure the movements of wrist and finger. This device could be used in clinical treatment to check whether the patient performs finger movements regularly or not.

10.3 Proposed method

MCI patients show some deviations in their basic activities in the early stages of AD. Typically the earliest stages of AD go unnoticed and patients with AD are diagnosed later during the moderate and severe stages, which are irreversible. Continuous monitoring of an MCI patient's daily basic activities could assist clinical experts in detecting AD at an early stage and providing proper treatment to slow its progression to the next stage. Some of the signs and symptoms of early-stage AD could be identified based on ADLs using inertial sensors. The most common symptoms appearing in the early stage of AD are as follows:

Forgetting basic activities: Those affected by AD can forget basic activities such as bathing, brushing teeth, etc. In some cases, they forget to eat and drink. However, the symptoms and signs of AD vary from person to person.

Repeating basic activity: AD patients frequently exhibit repetition of activities. For example, they may eat many times in a day, because they are unable to recall whether they have already eaten previously.

Confusion about date and time: AD patients often are confused as to date and time, so they might perform some activities at unusual times, for example, preparing food at midnight.

Duration: AD patients often take longer to accomplish an activity than a normal person would.

Currently, there is no dataset containing abnormal behavior of dementia patients using wearable sensors. In this work, a synthetic dataset has been generated using the University of California, Irvine (UCI) HMP dataset [5] by inserting abnormal activity.

The proposed system contains an activity recognition (AR) module and an abnormal activity detection (AAD) module. The AR module receives input from a wearable device and detects ADLs using ensemble-learning techniques. The AR module provides input to the AAD module in the form of the detected activity with a timestamp value. The AAD module identifies the abnormal behavior based on duration of activity, frequency of activity, and time of the activity using the timestamp value. The AAD module then classifies the detected activity into normal or abnormal activity.

For the purposes of this study, the deviations in daily activities have been generated artificially by inserting activities multiple times, skipping some activities, and inserting activities at unusual times. The tri-axis accelerometer generates data according to acceleration in the x-axis, y-axis, and z-axis directions with the timestamp value. The activities at unusual times, duration of activity, and repetitions of activity in the same day can be identified using the timestamp value. The wearable inertial sensor could track the basic activities such as brushing teeth, bathing, sleeping, eating, sitting, standing, and walking. The overall framework of the proposed system is illustrated in Fig. 10.1.

10.3.1 Activity recognition (AR) module

The wearable tri-axis accelerometer generates data with a timestamp value according to the ADL. The preprocessed data is sent to the AR module to detect the human activity. In the study described here, the HMP dataset is given as input to the AR module, which evaluates the sensor data using an ensemble of machine-learning techniques, such as bagging, boosting, and stacking. Experimentally, six activities, which are brushing teeth, combing hair, eating meat, drinking from a glass, walking, and lying in bed, are taken to train the AR module, as these are all basic activities carried out by elders in their daily lives. The detected activities and their timestamp details are fed into the AAD module to detect abnormal behavior.

10.3.2 Abnormal activity detection (AAD) module

The AAD module receives the input from the AR module and extracts the features including activity ID, date, time, duration, and frequency of activity, which are given in Table 10.2. These features are used to detect any anomalies in activities performed by MCI patients. The basic daily activities such as eating, sleeping, etc. can be different among elders, but some activities are definitely considered as abnormal. For example, walking for a long duration at midnight would be considered abnormal. In the proposed work, forgetting a basic activity, repeating an activity, showing confusion about date and time, and doing an activity for a long time are taken as abnormal activities, as these are common symptoms found in AD patients. For elderly patients with MCI, their daily activities are typically similar to or the same as those of non-MCI patients. The MCI patients only show abnormal deviations in their daily activities in the early stages of AD. In the proposed work, the historical daily activities of MCI patients are recorded and these activities are labeled as a "tolerable" level of activities.

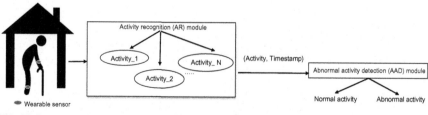

FIG. 10.1

Proposed system architecture for early detection of cognitive impairment.

Table 10.2 Extracted features from sensor data with timestamp.

Extracted features
• Activity ID
• Day of the activity
• Time of day
• Duration
• Frequency

The AAD module executes an abnormal activity detection algorithm, as shown in Algorithm 10.1. The historical activities of MCI patients are recorded in terms of day of the activity, frequency of the activity, duration of activity, and time of activity. The normal activities are labeled as a tolerable level of activity, as depicted in Algorithm 10.1. The abnormal activity detection algorithm detects abnormal behavior based on frequency of activity, duration for activity, and time of activity. As the proposed system evaluates the everyday activities, the health status of MCI patients could also be monitored continuously from a remote location.

Algorithm 10.1: Abnormal activity detection

```
Define Aᵢ: Recognized Activity with ID, i
Define i: Activity ID ranges from i= 1, 2....N
Define j: Day of the week ranges from j= 1, 2....7
Define k: Week of the month ranges from k=1, 2...4 or k = 1, 2, ... 5
Define Fᵢ: Frequency of activity Aᵢ
Define Dᵢ: Duration of activity Aᵢ
Define Tᵢ: Time of activity Aᵢ
Define Abᵢ: Abnormal activity of Aᵢ
Define Nᵢ: Normal activity of Aᵢ
Define T_Fi: Set of tolerable level of frequency of Aᵢ, {T_Fi}
Define T_Di: Set of tolerable level of duration of Aᵢ, {T_Di}
Define T_Ti: Set of tolerable level of time of Aᵢ, {T_Ti}
Define T_Si: Set of tolerable level of Skipping of activity Aᵢ,{T_Si}

Activity Recognition ( ):
Read tri-axis accelerometer sensor data;
Execute ensemble machine learning;
return (Activity_i, Timestamp);

Abnormal Activity Detection (Activity_i, Timestamp):
for every activity Aᵢ
        Compute (Fᵢ, Dᵢ, Tᵢ)
        Frequency_Aᵢ(Fᵢ);
        Duration_Aᵢ(Dᵢ);
        Time_Aᵢ(Tᵢ);
```

```
Frequency_A₁ (F₁):
if (F₁ >T_Fᵢ)
return (Ab₁)
else
            for all j
            Sum_week< - Sum (F₁)
            If (Sum_week>T_Fᵢ)
            return (Ab₁);
            else
            for all k
            Sum_month<- Sum (F₁)
                if (Sum_month>T_Fᵢ)
            return(Ab₁ )
            else
                    return(N₁)
if (F₁ ==0)
return (Ab₁)
else
            for all j
            Sum_week< - Sum (F₁)
              if (Sum_week ==0)
            return (Ab₁);

Duration_Ai (Di):
            if (D₁ >T_Dᵢ)
            return (Ab₁)
            else
            return(N₁)

Time_A₁(T₁):
for all j
Estimate T₁;
            if T₁ is not the element of { T_Tᵢ}
                    return (Ab₁)
```

10.4 Results and discussion

The AR module executes the ensemble machine-learning techniques and classifies the basic activities such as brushing teeth, combing hair, eating meat, drinking from a glass, walking, and lying in bed. The ensemble machine-learning techniques, which include SVM with a radial basis kernel function, random forest, C5.0, stochastic gradient boosting, bagged CART, and classification and regression trees (CART), are applied for training and testing the model; a10-fold cross-validation is used to evaluate the classification accuracy.

Out of six ensemble algorithms, the stochastic gradient boosting generates a better classification accuracy, 84.75%, than others, as depicted in Table 10.3 and Fig. 10.2. Random forest yields a classification accuracy of 84.09%, which is nearest

Table 10.3 Confusion matrix for ensemble techniques.

Ensemble machine-learning techniques	Activities of daily living	Brush_teeth	Comb_hair	Drink_glass	Eat_meat	Liedown_bed	Walk	Classification accuracy (%)
Support Vector Machine with a Radial Basis Kernel Function (SVM)	Brush_teeth	233	18	9	3	6	4	82.77
	Comb_hair	4	79	1	2	3	1	
	Drink_glass	9	4	97	4	6	5	
	Eat_meat	0	2	5	211	7	1	
	Liedown_bed	3	5	0	0	52	6	
	Walk	3	58	0	0	14	207	
Random Forest	Brush_teeth	222	16	13	3	5	4	84.09
	Comb_hair	7	120	1	2	4	24	
	Drink_glass	13	5	95	5	4	3	
	Eat_meat	1	2	3	209	3	0	
	Liedown_bed	5	3	0	1	62	8	
	Walk	4	20	0	0	10	185	
C5.0	Brush_teeth	221	19	12	2	5	6	83.5
	Comb_hair	6	116	2	3	3	24	
	Drink_glass	14	4	93	4	3	2	
	Eat_meat	1	3	4	209	4	1	
	Liedown_bed	6	4	1	2	62	5	
	Walk	4	20	0	0	11	186	
Stochastic Gradient Boosting	Brush_teeth	230	18	9	2	4	4	84.75
	Comb_hair	5	125	1	5	9	26	
	Drink_glass	10	3	95	4	5	1	
	Eat_meat	0	4	5	209	4	3	
	Liedown_bed	6	3	2	0	56	5	
	Walk	1	13	0	0	10	185	

Bagged CART	Brush_teeth	Comb_hair	Drink_glass	Eat_meat	Liedown_bed	Walk	83.43
Brush_teeth	223	18	13	4	6	4	
Comb_hair	7	120	3	2	4	25	
Drink_glass	13	4	93	5	3	3	
Eat_meat	1	2	3	207	4	1	
Liedown_bed	6	4	0	2	61	9	
Walk	2	18	0	0	10	182	
Classification and Regression Trees (CART)							60.73
Brush_teeth	228	68	13	11	10	9	
Comb_hair	0	0	0	0	0	0	
Drink_glass	0	0	0	0	0	0	
Eat_meat	12	37	99	209	56	7	
Liedown_bed	0	0	0	0	0	0	
Walk	12	61	0	0	22	208	

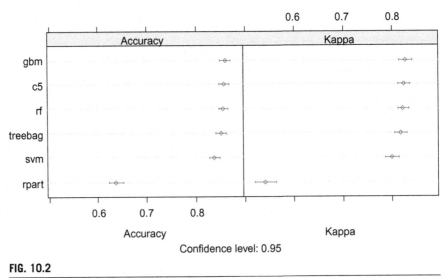

FIG. 10.2

Classification accuracy of ensemble techniques.

to the stochastic gradient boosting algorithm. The confusion matrix obtained using various ensemble techniques is illustrated in Table 10.3.

The detected activity with its timestamp value sent to the AAD module is processed using the abnormal activity detection algorithm, which is depicted in Algorithm 10.1. The proposed abnormal activity detection algorithm detects the abnormal and normal behavior of the MCI patient, as illustrated in Figs. 10.3A and B, 10.4, and 10.5. The proposed algorithm identifies the normal activity of the MCI patient based on frequency of activity performed in a day, as shown in Fig. 10.3A.

The abnormal activity of the MCI patient in terms of frequency is shown in Fig. 10.3B. An AD patient typically has confusion about date and time. Experimentally, the activity of brushing teeth is singled out, as illustrated in Fig. 10.4, which shows the time of activity between MCI patients and AD patients. At an early stage of AD, the patient spends a lot of time completing every activity, which is illustrated in Fig. 10.5. However, the signs and symptoms of AD can differ from person to person. The progression of AD is also different among patients. This work considers the common signs and symptoms identified in AD patients to evaluate the early stage of AD.

The proposed system uses a synthetic dataset to detect abnormal behavior of MCI patients. In future, a real-time dataset will be used to evaluate the ADLs of MCI patients. The different signs and symptoms of AD will be included in future to assess abnormal activities.

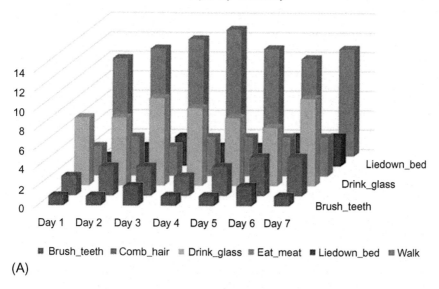

(A)

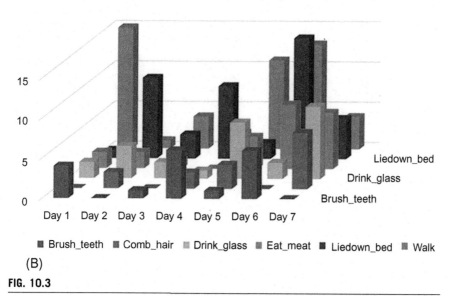

(B)

FIG. 10.3

(A) Frequency of activities at MCI stage. (B) Frequency of activities at AD stage.

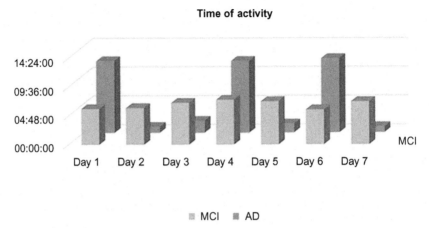

FIG. 10.4

Time of brushing teeth between MCI and AD.

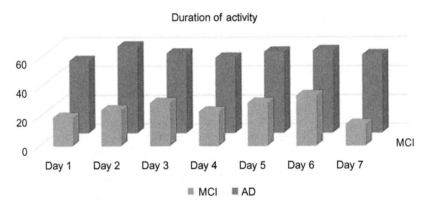

FIG. 10.5

Duration of eating between MCI and AD.

10.5 Conclusion

This chapter presents a model for predicting AD disease at an early stage based on ADLs. The proposed system uses a wearable device, which continuously monitors the basic activities of MCI patients in their daily life. Based on the recorded historical daily activities of MCI patients, any deviations in ADLs are detected. The timestamp values generated along with the sensor data are the most important factors to differentiate an abnormal activity from a normal activity. The ensemble machine-learning techniques detect the basic activities with better classification accuracy. The proposed abnormal activity detection algorithm detects normal and abnormal behavior based on frequency of activity, time of activity, and duration of activity. In future, the proposed work will be implemented using real-time patient daily activities. Moreover, the abnormal behavior of MCI patients will be identified at a high accuracy by considering other signs and symptoms.

References

[1] A.M. Rahmani, et al., Exploiting smart E-health gateways at the edge of healthcare internet-of-things: a fog computing approach, Futur. Gener. Comput. Syst. 78 (2) (2018) 641–658.

[2] A. Bulling, U. Blanke, B. Schiele, A tutorial on human activity recognition using body-worn inertial sensors, ACM Comput. Surv. (CSUR) 46 (3) (2014).

[3] Mild Cognitive Impairment (MCI), https:/www.alz.org/alzheimers-dementia/what-is-dementia/related_conditions/mild-cognitive-impairment.

[4] Mild Cognitive Impairment (MCI), https:/www.alzheimers.org.uk/about-dementia/types-dementia/mild-cognitive-impairment-mci.

[5] D. Dua, C. Graff, UCI Machine Learning Repository, University of California, School of Information and Computer Science, Irvine, CA, 2019.http:/archive.ics.uci.edu/ml.

[6] R. Varatharajan, G. Manogaran, M.K. Priyan, R. Sundarasekar, Wearable sensor devices for early detection of Alzheimer disease using dynamic time warping algorithm, Clust. Comput. (2017) 1–10.

[7] S. Pedro, J. Quintas, P. Menezes, Sensor-based detection of Alzheimer's disease-related behaviors, in: The International Conference on Health Informatics, 2014, pp. 276–279.

[8] K.S. Gayathri, K.S. Easwarakumar, Intelligent decision support system for dementia care through smart home, Proc. Comput. Sci. 93 (2016) 947–955.

[9] A. Lotfi, C. Langensiepen, S.M. Mahmoud, M.J. Akhlaghinia, Smart homes for the elderly dementia sufferers: identification and prediction of abnormal behaviour, J. Ambient. Intell. Humaniz. Comput. 3 (3) (2012) 205–218.

[10] Y.-L. Hsu, et al., Gait and balance analysis for patients with Alzheimer's disease using an inertial-sensor-based wearable instrument, IEEE J. Biomed. Health Inform. 18 (6) (2014) 1822–1830.

[11] M. Derawi, P. Bours, Gait and activity recognition using commercial phones, Comput. Secur. 39 (2013) 137–144.

[12] D. De, P. Bharti, S.K. Das, S. Chellappan, Multimodal wearable sensing for fine-grained activity recognition in healthcare, IEEE Internet Comput. 19 (5) (2015) 26–35.

[13] G. Panahandeh, N. Mohammadiha, A. Leijon, P. Händel, Continuous hidden Markov model for pedestrian activity classification and gait analysis, IEEE Trans. Instrum. Meas. 62 (5) (2013) 1073–1083.

[14] A.M. Cristiani, G.M. Bertolotti, E. Marenzi, S. Ramat, An instrumented insole for long term monitoring movement, comfort, and ergonomics, IEEE Sensors J. 14 (5) (2014) 1564–1572.

[15] N. Friedman, J.B. Rowe, D.J. Reinkensmeyer, M. Bachman, The manumeter: a wearable device for monitoring daily use of the wrist and fingers, IEEE J. Biomed. Health Inform. 18 (6) (2014) 1804–1812.

Index

Note: Page numbers followed by *f* indicate figures, *t* indicate tables, and *b* indicate boxes.

Printed in the United States
By Bookmasters